A Little Tour in France

Henry James

Printed in Scotts Valley, CA – USA.

James, Henry.

A Little Tour in France / Henry James – 1st ed.

1. Travel

Book Cover Photo:

© Diademimages | Dreamstime.com

We good Americans - I say it without presumption - are too apt to think that France is Paris, just as we are accused of being too apt to think that Paris is the celestial city. This is by no means the case, fortunately for those persons who take an interest in modern Gaul, and yet are still left vaguely unsatisfied by that epitome of civilization which stretches from the Arc de Triomphe to the Gymnase theatre. It had already been intimated to the author of these light pages that there are many good things in the *doux pays de France* of which you get no hint in a walk between those ornaments of the capital; but the truth had been revealed only in quick-flashing glimpses, and he was conscious of a desire to look it well in the face. To this end he started, one rainy morning in mid-September, for the charming little city of Tours, from which point it seemed possible to make a variety of fruitful excursions. His excursions resolved themselves ultimately into a journey through several provinces, a journey which had its dull moments (as one may defy any journey not to have), but which enabled him to feel that his proposition was demonstrated. France may be Paris, but Paris is not France; that was perfectly evident on the return to the capital.

I must not speak, however, as if I had discovered the provinces. They were discovered, or at least revealed by Balzac, if by any one, and are now easily accessible to visitors. It is true, I met no visitors, or only one or two, whom it was pleasant to meet. Throughout my little tour I was almost the only tourist. That is perhaps one reason why it was so successful.

1

I am ashamed to begin with saying that Touraine is the garden of France; that remark has long ago lost its bloom. The town of Tours, however, has some thing sweet and bright, which suggests that it is surrounded by a land of fruits. It is a very agreeable little city; few towns of its size are more ripe, more complete, or, I should suppose, in better humor with themselves and less disposed to envy the responsibilities of bigger places. It is truly the capital of its smiling province; a region of easy abundance, of good living, of genial, comfortable, optimistic, rather indolent opinions. Balzac says in one of his tales that the real Tourangeau will not make an effort, or displace himself even, to go in search of a pleasure; and it is not difficult to understand the sources of this amiable cynicism. He must have a vague conviction that he can only lose by almost any change. Fortune has been kind to him: he lives in a temperate, reasonable, sociable climate, on the banks, of a river which, it is true, sometimes floods the country around it, but of which the ravages appear to be so easily repaired that its aggressions may perhaps be regarded (in a region where so many good things are certain) merely as an occasion for healthy suspense. He is surrounded by fine old traditions, religious, social, architectural, culinary; and he may have the satisfaction of feeling that he is French to the core. No part of his admirable country is more characteristically national. Normandy is Normandy, Burgundy is Burgundy, Provence is Provence; but Touraine is essentially France. It is the land of Rabelais, of Descartes, of Balzac, of good books and good company, as well as good dinners and good houses. George Sand has somewhere a charming passage about the mildness, the convenient quality, of the physical conditions of central France, "son climat souple et chaud, ses pluies abondantes et courtes." In the autumn of 1882 the rains perhaps were less short than abundant; but when the days were fine it was impossible that anything in the way of weather could be more charming. The vineyards and orchards looked rich in the fresh, gay light; cultivation was everywhere, but everywhere it seemed to be easy. There was no visible poverty; thrift and success presented themselves as matters of good taste. The white caps of the women glittered in the sunshine, and their well-made sabots clicked cheerfully on the hard, clean roads. Touraine is a land of old chateaux, a gallery of architectural specimens and of large hereditary properties. The peasantry have less of the luxury of ownership than in most other parts of France; though they have enough of it to give them quite their share of that shrewdly conservative look which, in the little, chaffering, *place* of the market-town, the stranger observes so often in the wrinkled brown masks that surmount the agricultural blouse. This is, moreover, the heart of the old French monarchy; and as that monarchy was splendid and picturesque, a reflection of the splendor still glitters in the current of the Loire. Some of the most striking events of French history have occurred on the banks of that river, and the soil it waters bloomed for a while with the flowering of the Renaissance. The Loire gives a great "style" to a landscape of which the features are not, as the phrase is, prominent, and carries the eye to distances even more poetic than the green horizons of Touraine. It is a very fitful stream, and is sometimes observed to run thin and

4

expose all the crudities of its channel, a great defect certainly in a river which is so much depended upon to give an air to the places it waters. But I speak of it as I saw it last; full, tranquil, powerful, bending in large slow curves, and sending back half the light of the sky. Nothing can be finer than the view of its course which you get from the battlements and terraces of Amboise. As I looked down on it from that elevation one lovely Sunday morning, through a mild glitter of autumn sunshine, it seemed the very model of a generous, beneficent stream. The most charming part of Tours is naturally the shaded quay that overlooks it, and looks across too at the friendly faubourg of Saint Symphorien and at the terraced heights which rise above this. Indeed, throughout Touraine, it is half the charm of the Loire that you can travel beside it. The great dike which protects it, or, protects the country from it, from Blois to Angers, is an admirable road; and on the other side, as well, the highway constantly keeps it company. A wide river, as you follow a wide road, is excellent company; it heightens and shortens the way.

The inns at Tours are in another quarter, and one of them, which is midway between the town and the station, is very good. It is worth mentioning for the fact that every one belonging to it is extraordinarily polite, so unnaturally polite as at first to excite your suspicion that the hotel has some hidden vice, so that the waiters and chambermaids are trying to pacify you in advance. There was one waiter in especial who was the most accomplished social being I have ever encountered; from morning till night he kept up an inarticulate murmur of urbanity, like the hum of a spinning-top. I may add that I discovered no dark secrets at the Hotel de l'Univers; for it is not a secret to any traveler to-day that the obligation to partake of a lukewarm dinner in an overheated room is as imperative as it is detestable. For the rest, at Tours, there is a certain Rue Royale which has pretensions to the monumental; it was constructed a hundred years ago, and the houses, all alike, have on a moderate scale a pompous eighteenth-century look. It connects the Palais de Justice, the most important secular building in the town, with the long bridge which spans the Loire, the spacious, solid bridge pronounced by Balzac, in "Le Cure de Tours," "one of the finest monuments of French architecture." The Palais de Justice was the seat of the Government of Leon Gambetta in the autumn of 1870, after the dictator had been obliged to retire in his balloon from Paris, and before the Assembly was constituted at Bordeaux. The Germans occupied Tours during that terrible winter; it is astonishing, the number of places the Germans occupied. It is hardly too much to say that wherever one goes in, certain parts of France, one encounters two great historic facts: one is the Revolution; the other is the German invasion. The traces of the Revolution remain in a hundred scars and bruises and mutilations, but the visible marks of the war of 1870 have passed away. The country is so rich, so living, that she has been able to dress her wounds, to hold up her head, to smile again; so that the shadow of that darkness has ceased to rest upon her. But what you do not see you still may hear; and one remembers with a certain shudder that only a few short years ago this province, so intimately French, was under the heel of a foreign foe. To be intimately French was apparently not a safeguard; for so successful an invader it could only be a challenge. Peace and plenty, however, have succeeded that episode; and among the gardens and vineyards of Touraine it seems, only a legend the more in a country of legends.

It was not, all the same, for the sake of this checkered story that I mentioned the Palais de Justice and the Rue Royale. The most interesting fact, to my mind, about the high-street of Tours was that as you walked toward the bridge on the right-hand *trottoir* you can look up at the house, on the other side of the way, in which Honore de Balzac first saw the light. That violent and complicated genius was a child of the good-humored and succulent Touraine. There is something anomalous in the fact, though, if one thinks about it a little, one may discover certain correspondences between his character and that of his native province. Strenuous, laborious, constantly in felicitous in spite of his great successes, he suggests at times a very different set of influences. But he had his jovial, full-feeding side, the side that comes out in the "Contes Drolatiques," which are the romantic and epicurean chronicle of the old manors and abbeys of this region. And he was, moreover, the product of a soil into which a great deal of history had been trodden. Balzac was genuinely as well as affectedly monarchical, and he was saturated with, a sense of the past. Number 39 Rue Royale of which the basement, like all the basements in the Rue Royale, is occupied by a shop is not shown to the public; and I know not whether tradition designates the chamber in which the author of "Le Lys dans la Vallee" opened his eyes into a world in which he was to see and to imagine such extraordinary things. If this were the case, I would willingly have crossed its threshold; not for the sake of any relic of the great novelist which it may possibly contain, nor even for that of any mystic virtue which may be supposed to reside within its walls, but simply because to look at those four modest walls can hardly fail to give one a strong impression of the force of human endeavor. Balzac, in the maturity of his vision, took in more of human life than any one, since Shakespeare, who has attempted to tell us stories about it; and the very small scene on which his consciousness dawned is one end of the immense scale that he traversed. I confess it shocked me a little to find that he was born in a house "in a row," a house, moreover, which at the date of his birth must have been only about twenty years old. All that is contradictory. If the tenement selected for this honor could not be ancient and embrowned, it should at least have been detached.

There is a charming description, in his little tale of "La Grenadiere," of the view of the opposite side of the Loire as you have it from the square at the end of the Rue Royale, a square that has some pretensions to grandeur, overlooked as it is by the Hotel de Ville and the Musee, a pair of edifices which directly contemplate the river, and ornamented with marble images of Francois Rabelais and Rene Descartes. The former, erected a few years since, is a very honorable production; the pedestal of the latter could, as a matter of course, only be inscribed with the *Cogito ergo Sum.* The two statues mark the two opposite poles to which the brilliant French mind has traveled; and if there were an effigy of Balzac at Tours, it ought to stand midway between them. Not that he, by any means always struck the happy mean between the sensible and the metaphysical; but one may say of him that half of his genius looks in one direction and half in the other. The side that turns toward Francois Rabelais would be, on the whole, the side that takes the sun. But there is no statue of Balzac at Tours; there is only, in one of the chambers of the melancholy museum, a rather clever, coarse bust. The description in "La Grenadiere," of which I just spoke, is too long to quote; neither have I space for any one of the brilliant attempts at landscape painting which are woven into the shimmering texture of "Le Lys dans la Vallee." The little manor of Clochegourde, the residence of Madame de Mortsauf, the heroine of that extraordinary work, was within a moderate walk of Tours, and the picture in the novel is presumably a copy from an original which it would be possible to-day to discover. I did not, however, even make the attempt. There are so many chateaux in Touraine commemorated in history, that it would take one too far to look up those which have been commemorated in fiction. The most I did was to endeavor to identify the former residence of Mademoiselle Gamard, the sinister old maid of "Le Cure de Tours." This terrible woman occupied a small house in the rear of the cathedral, where I spent a whole morning in wondering rather stupidly which house it could be. To reach the cathedral from the little *place* where we stopped just now to look across at the Grenadiere, without, it must be confessed, very vividly seeing it, you follow the quay to the right, and pass out of sight of the charming *coteau* which, from beyond the river, faces the town, a soft agglomeration of gardens, vineyards, scattered villas, gables and turrets of slateroofed chateaux, terraces with gray balustrades, mossgrown walls draped in scarlet Virginia-creeper. You turn into the town again beside a great military barrack which is ornamented with a rugged mediaeval tower, a relic of the ancient fortifications, known to the Tourangeaux of to-day as the Tour de Guise. The young Prince of Joinville, son of that Duke of Guise who was murdered by the order of Henry II. at Blois, was, after the death of his father, confined here for more than two years, but made his escape one summer evening in 1591, under the nose of his keepers, with a gallant audacity which has attached the memory of the exploit to his sullen-looking prison. Tours has a garrison of five regiments, and the little red-legged soldiers light up the town. You see them stroll upon the clean, non-commercial quay, where there are no signs of navigation, not even by oar, no barrels nor bales, no loading nor unloading, no masts against the sky nor booming of steam in the air. The most active business that goes on there is that patient and fruitless angling in, which the French, as the votaries of art for art, excel all other people. The little soldiers, weighed down by the contents of their enormous pockets, pass with respect from one of these masters of the rod to the other, as he sits soaking an indefinite bait in the large, indifferent stream. After you turn your back to the quay you have only to go a little way before you reach the cathedral.

2

It is a very beautiful church of the second order of importance, with a charming mouse-colored complexion and a pair of fantastic towers. There is a commodious little square in front of it, from which you may look up at its very ornamental face; but for purposes of frank admiration the sides and the rear are perhaps not sufficiently detached. The cathedral of Tours, which is dedicated to Saint Gatianus, took a long time to build. Begun in 1170, it was finished only in the first half of the sixteenth century; but the ages and the weather have interfused so well the tone of the different parts, that it presents, at first at least, no striking incongruities, and looks even exceptionally harmonious and complete. There are many grander cathedrals, but there are probably few more pleasing; and this effect of delicacy and grace is at its best toward the close of a quiet afternoon, when the densely decorated towers, rising above the little Place de l'Archeveche, lift their curious lanterns into the slanting light, and offer a multitudinous perch to troops of circling pigeons. The whole front, at such a time, has an appearance of great richness, although the niches which surround the three high doors (with recesses deep enough for several circles of sculpture) and indent the four great buttresses that ascend beside the huge rose-window, carry no figures beneath their little chiseled canopies. The blast of the great Revolution blew down most of the statues in France, and the wind has never set very strongly toward putting them up again. The embossed and crocketed cupolas which crown the towers of Saint Gatien are not very pure in taste; but, like a good many impurities, they have a certain character. The interior has a stately slimness with which no fault is to be found, and which in the choir, rich in early glass and surrounded by a broad passage, becomes very bold and noble. Its principal treasure, perhaps, is the charming little tomb of the two children (who died young) of Charles VIII. and Anne of Brittany, in white marble, embossed with symbolic dolphins and exquisite arabesques. The little boy and girl lie side by side on a slab of black marble, and a pair of small kneeling angels, both at their head and at their feet, watch over them. Nothing could be more perfect than this monument, which is the work of Michel Colomb, one of the earlier glories of the French Renaissance; it is really a lesson in good taste. Originally placed in the great abbey-church of Saint Martin, which was for so many ages the holy place of Tours, it happily survived the devastation to which that edifice, already sadly shattered by the wars of religion and successive profanations, finally succumbed in 1797. In 1815 the tomb found an asylum in a quiet corner of the cathedral.

I ought, perhaps, to be ashamed to acknowledge, that I found the profane name of Balzac capable of adding an interest even to this venerable sanctuary. Those who have read the terrible little story of "Le Cure de Tours" will perhaps remember that, as I have already mentioned, the simple and childlike old Abbe Birotteau, victim of the infernal machinations of the Abbe Troubert and Mademoiselle Gamard, had his quarters in the house of that lady (she had a specialty of letting lodgings to priests), which stood on the north side of the cathedral, so close under its walls that the supporting pillar of one of the great flying buttresses was planted in the spinster's garden. If you wander round behind the church, in search of this more than historic habitation, you will have occasion to see that the side and rear of Saint Gatien make a delectable and curious figure. A narrow lane passes beside the high wall which conceals from sight the palace of the archbishop, and beneath the flying buttresses, the far-projecting gargoyles, and the fine south porch of the church. It terminates in a little, dead, grass-grown square entitled the Place Gregoire de Tours. All this part of the exterior of the cathedral is very brown, ancient, Gothic, grotesque; Balzac calls the whole place "a desert of stone." A battered and gabled wing, or out-house (as it appears to be) of the hidden palace, with a queer old stone pulpit jutting out from it, looks down on this melancholy spot, on the other side of which is a seminary for young priests, one of whom issues from a door in a quiet corner, and, holding it open a moment behind him, shows a glimpse of a sunny garden, where you may fancy other black young figures strolling up and down. Mademoiselle Gamard's house, where she took her two abbes to board, and basely conspired with one against the other, is still further round the cathedral. You cannot quite put your hand upon it today, for the dwelling which you say to yourself that it *must* have been Mademoiselle Gamard's does not fulfill all the conditions mentioned in Balzac's description. The edifice in question, however, fulfils conditions enough; in particular, its little court offers hospitality to the big buttress of the church. Another buttress, corresponding with this (the two, between them, sustain the gable of the north transept), is planted in the small cloister, of which the door on the further side of the little soundless Rue de la Psalette, where nothing seems ever to pass, opens opposite to that of Mademoiselle Gamard. There is a very genial old sacristan, who introduced me to this cloister from the church. It is very small and solitary, and much mutilated; but it nestles with a kind of wasted friendliness beneath the big walls of the cathedral. Its lower arcades have been closed, and it has a small plot of garden in the middle, with fruit-trees which I should imagine to be too much overshadowed. In one corner is a remarkably picturesque turret, the cage of a winding staircase which ascends (no great distance) to an upper gallery, where an old priest, the *chanoine-gardien* of the church, was walking to and fro with his breviary. The turret, the gallery, and even the chanoine-gardien, belonged, that sweet September morning, to the class of objects that are dear to painters in water-colors.

3

I have mentioned the church of Saint Martin, which was for many years the sacred spot, the shrine of pilgrimage, of Tours. Originally the simple burial place of the great apostle who in the fourth century Christianized Gaul, and who, in his day a brilliant missionary and worker of miracles, is chiefly known to modem fame as the worthy that cut his cloak in two at the gate of Amiens to share it with a beggar (tradition fails to say, I believe, what he did with the other half), the abbey of Saint Martin, through the Middle Ages, waxed rich and powerful, till it was known at last as one of the most luxurious religious houses in Christendom, with kings for its titular abbots (who, like Francis I., sometimes turned and despoiled it) and a great treasure of precious things. It passed, however, through many vicissitudes. Pillaged by the Normans in the ninth century and by the Huguenots in the sixteenth, it received its death-blow from the Revolution, which must have brought to bear upon it an energy of destruction proportionate to its mighty bulk. At the end of the last century a huge group of ruins alone remained, and what we see to-day may be called the ruin of a ruin. It is difficult to understand how so vast an edifice can have been so completely obliterated. Its site is given up to several ugly streets, and a pair of tall towers, separated by a space which speaks volumes as to the size of the church, and looking across the close-pressed roofs to the happier spires of the cathedral, preserved for the modern world the memory of a great fortune, a great abuse, perhaps, and at all events a great penalty. One may believe that to this day a considerable part of the foundations of the great abbey is buried in the soil of Tours. The two surviving towers, which are dissimilar in shape, are enormous; with those of the cathedral they form the great landmarks of the town. One of them bears the name of the Tour de l'Horloge; the other, the so-called Tour Charlemagne, was erected (two centuries after her death) over the tomb of Luitgarde, wife of the great Emperor, who died at Tours in 800. I do not pretend to understand in what relation these very mighty and effectually detached masses of masonry stood to each other, but in their gray elevation and loneliness they are striking and suggestive to-day; holding their hoary heads far above the modern life of the town, and looking sad and conscious, as they had outlived all uses. I know not what is supposed to have become of the bones of the blessed saint during the various scenes of confusion in which they may have got mislaid; but a mystic connection with his wonder-working relics may be perceived in a strange little sanctuary on the left of the street, which opens in front of the Tour Charlemagne, the rugged base of which, by the way, inhabited like a cave, with a diminutive doorway, in which, as I passed, an old woman stood cleaning a pot, and a little dark window decorated with homely flowers, would be appreciated by a painter in search of "bits." The present shrine of Saint Martin is enclosed (provisionally, I suppose) in a very modem structure of timber, where in a dusky cellar, to which you descend by a wooden staircase adorned with votive tablets and paper roses, is placed a tabernacle surrounded by twinkling tapers and prostrate worshippers. Even this crepuscular vault, however, fails, I think, to attain solemnity; for the whole place is strangely vulgar and garish. The Catholic church, as churches go to-day, is certainly the most spectacular; but it must feel that it has a great fund of impressiveness to draw upon when it opens such sordid little shops of sanctity as this. It is impossible not to be struck with the grotesqueness of such an establishment, as the last link in the chain of a great ecclesiastical tradition.

In the same street, on the other side, a little below, is something better worth your visit than the shrine of Saint Martin. Knock at a high door in a white wall (there is a cross above it), and a fresh-faced sister of the convent of the Petit Saint Martin will let you into the charming little cloister, or rather fragment of a cloister. Only one side of this exquisite structure remains, but the whole place is effective. In front of the beautiful arcade, which is terribly bruised and obliterated, is one of those walks of interlaced *tilleuls* which are so frequent in Touraine, and into which the green light filters so softly through a lattice of clipped twigs. Beyond this is a garden, and beyond the garden are the other buildings of the Convent, where the placid sisters keep a school, a test, doubtless, of placidity. The imperfect arcade, which dates from the beginning of the sixteenth century (I know nothing of it but what is related in Mrs. Pattison's "Renaissance in France") is a truly enchanting piece of work; the cornice and the angles of the arches, being covered with the daintiest sculpture of arabesques, flowers, fruit, medallions, cherubs, griffins, all in the finest and most attenuated relief. It is like the chasing of a bracelet in stone. The taste, the fancy, the elegance, the refinement, are of those things which revive our standard of the exquisite. Such a piece of work is the purest flower of the French Renaissance; there is nothing more delicate in all Touraine.

There is another fine thing at Tours which is not particularly delicate, but which makes a great impression, the very interesting old church of Saint Julian, lurking in a crooked corner at the right of the Rue Royale, near the point at which this indifferent thoroughfare emerges, with its little cry of admiration, on the bank of the Loire. Saint Julian stands to-day in a kind of neglected hollow, where it is much shut in by houses; but in the year 1225, when the edifice was begun, the site was doubtless, as the architects say, more eligible. At present, indeed, when once you have caught a glimpse of the stout, serious Romanesque tower, which is not high, but strong, you feel that the building has something to say, and that you must stop to listen to it. Within, it has a vast and splendid nave, of immense height, the nave of a cathedral, with a shallow choir and transepts, and some admirable old glass. I spent half an hour there one morning, listening to what the church had to say, in perfect solitude. Not a worshipper entered, not even an old man with a broom. I have always thought there is a sex in fine buildings; and Saint Julian, with its noble nave, is of the gender of the name of its patron.

It was that same morning, I think, that I went in search of the old houses of Tours; for the town contains several goodly specimens of the domestic architecture of the past. The dwelling to which the average Anglo-Saxon will most promptly direct his steps, and the only one I have space to mention, is the so-called Maison de Tristan l'Hermite, a gentleman whom the readers of "Quentin Durward" will not have forgotten, the hangman-in-ordinary to the great King Louis XI. Unfortunately the house of Tristan is not the house of Tristan at all; this illusion has been cruelly dispelled. There are no illusions left, at all, in the good city of Tours, with regard to Louis XI. His terrible castle of Plessis, the picture of which sends a shiver through the youthful reader of Scott, has been reduced to suburban insignificance; and the residence of his *triste compere,* on the front of which a festooned rope figures as a motive for decoration, is observed to have been erected in the succeeding century. The Maison de Tristan may be visited for itself, however, if not for Walter Scott; it is an exceedingly picturesque old facade, to which you pick your way through a narrow and tortuous street, a street terminating, a little beyond it, in the walk beside the river. An elegant Gothic doorway is let into the rusty-red brick-work, and strange little beasts crouch at the angles of the windows, which are surmounted by a tall graduated gable, pierced with a small orifice, where the large surface of brick, lifted out of the shadow of the street, looks yellow and faded. The whole thing is disfigured and decayed; but it is a capital subject for a sketch in colors. Only I must wish the sketcher better luck or a better temper than my own. If he ring the bell to be admitted to see the court, which I believe is more sketchable still, let him have patience to wait till the bell is answered. He can do the outside while they are coming.

The Maison de Tristan, I say, may be visited for itself; but I hardly know what the remnants of Plessisles-Tours may be visited for. To reach them you wander through crooked suburban lanes, down the course of the Loire, to a rough, undesirable, incongruous spot, where a small, crude building of red brick is pointed out to you by your cabman (if you happen to drive) as the romantic abode of a superstitious king, and where a strong odor of pigsties and other unclean things so prostrates you for the moment that you have no energy to protest against the obvious fiction. You enter a yard encumbered with rubbish and a defiant dog, and an old woman emerges from a shabby lodge and assures you that you are indeed in an historic place. The red brick building, which looks like a small factory, rises on the ruins of the favorite residence of the dreadful Louis. It is now occupied by a company of night-scavengers, whose huge carts are drawn up in a row before it. I know not whether this be what is called the irony of fate; at any rate, the effect of it is to accentuate strongly the fact (and through the most susceptible of our senses) that there is no honor for the authors of great wrongs. The dreadful Louis is reduced simply to an offence to the nostrils. The old woman shows you a few fragments, several dark, damp, much-encumbered vaults, denominated dungeons, and an old tower staircase, in good condition. There are the outlines of the old moat; there is also the outline of the old guard-room, which is now a stable; and there are other vague outlines and inconsequent lumps, which I have forgotten. You need all your imagination, and even then you cannot make out that Plessis was a castle of large extent, though the old woman, as your eye wanders over the neighboring *potagers,* talks a good deal about the gardens and the park. The place looks mean and flat; and as you drive away you scarcely know whether to be glad or sorry that all those bristling horrors have been reduced to the commonplace.

A certain flatness of impression awaits you also, I think, at Marmoutier, which is the other indispensable excursion in the near neighborhood of Tours. The remains of this famous abbey lie on the other bank of the stream, about a mile and a half from the town. You

14

follow the edge of the big brown river; of a fine afternoon you will be glad to go further still. The abbey has gone the way of most abbeys; but the place is a restoration as well as a ruin, inasmuch as the sisters of the Sacred Heart have erected a terribly modern convent here. A large Gothic doorway, in a high fragment of ancient wall, admits you to a gardenlike enclosure, of great extent, from which you are further introduced into an extraordinarily tidy little parlor, where two good nuns sit at work. One of these came out with me, and showed me over the place, a very definite little woman, with pointed features, an intensely distinct enunciation, and those pretty manners which (for whatever other teachings it may be responsible) the Catholic church so often instills into its functionaries. I have never seen a woman who had got her lesson better than this little trotting, murmuring, edifying nun. The interest, of Marmoutier to-day is not so much an interest of vision, so to speak, as an interest of reflection, that is, if you choose to reflect (for instance) upon the wondrous legend of the seven sleepers (you may see where they lie in a row), who lived together they were brothers and cousins in primitive piety, in the sanctuary constructed by the blessed Saint Martin (emulous of his precursor, Saint Gatianus), in the face of the hillside that overhung the Loire, and who, twenty-five years after his death, yielded up their seven souls at the same moment, and enjoyed the curious privilege of retaining in their faces, in spite of this process, the rosy tints of life. The abbey of Marmoutier, which sprung from the grottos in the cliff to which Saint Gatianus and Saint Martin retired to pray, was therefore the creation of the latter worthy, as the other great abbey, in the town proper, was the monument of his repose. The cliff is still there; and a winding staircase, in the latest taste, enables you conveniently to explore its recesses. These sacred niches are scooped out of the rock, and will give you an impression if you cannot do without one. You will feel them to be sufficiently venerable when you learn that the particular pigeon-hole of Saint Gatianus, the first Christian missionary to Gaul, dates from the third century. They have been dealt with as the Catholic church deals with most of such places today; polished and furnished up; labeled and ticketed, *edited,* with notes, in short, like an old book. The process is a mistake, the early editions had more sanctity. The modern buildings (of the Sacred Heart), on which you look down from these points of vantage, are in the vulgar taste which seems doomed to stamp itself on all new Catholic work; but there was nevertheless a great sweetness in the scene. The afternoon was lovely, and it was flushing to a close. The large garden stretched beneath us, blooming with fruit and wine and succulent vegetables, and beyond it flowed

15

the shining river. The air was still, the shadows were long, and the place, after all, was full of memories, most of which might pass for virtuous. It certainly was better than Plessis-les-Tours.

4

Your business at Tours is to make excursions; and if you make them all, you will be very well occupied. Touraine is rich in antiquities; and an hour's drive from the town in almost any direction will bring you to the knowledge of some curious fragment of domestic or ecclesiastical architecture, some turreted manor, some lonely tower, some gabled village, or historic site. Even, however, if you do everything, which was not my case, you cannot hope to relate everything, and, fortunately for you, the excursions divide themselves into the greater and the less. You may achieve most of the greater in a week or two; but a summer in Touraine (which, by the way must be a charming thing) would contain none too many days for the others. If you come down to Tours from Paris, your best economy is to spend a few days at Blois, where a clumsy, but rather attractive little inn, on the edge of the river, will offer you a certain amount of that familiar and intermittent hospitality which a few weeks spent in the French provinces teaches you to regard as the highest attainable form of accommodation. Such an economy I was unable to practice. I could only go to Blois (from Tours) to spend the day; but this feat I accomplished twice over. It is a very sympathetic little town, as we say nowadays, and one might easily resign one's self to a week there. Seated on the north bank of the Loire, it presents a bright, clean face to the sun, and has that aspect of cheerful leisure which belongs to all white towns that reflect, themselves in shining waters. It is the water-front only of Blois, however, that exhibits, this fresh complexion; the interior is of a proper brownness, as befits a signally historic city. The only disappointment I had there was the discovery that the castle, which is the special object of one's pilgrimage, does not overhang the river, as I had always allowed myself to understand. It overhangs the town, but it is scarcely visible from the stream. That peculiar good fortune is reserved for Amboise and Chaurnont.

The Chateau de Blois is one of the most beautiful and elaborate of all the old royal residences of this part of France, and I suppose it should have all the honors of my description. As you cross its threshold, you step straight into the brilliant movement of the French Renaissance. But it is too rich to describe, I can only touch it here and there. It must be premised that in speaking of it as one sees it to-day, one speaks of a monument unsparingly restored. The work of restoration has been as ingenious as it is profuse, but it rather chills the imagination. This is perhaps almost the first thing you feel as you approach the castle from the streets of the town. These little streets, as they, leave the river, have pretensions to romantic steepness; one of them, indeed, which resolves itself into a high staircase with divergent wings (the *escalier monumental*), achieved this result so successfully as to remind me vaguely I hardly know why of the great slope of the Capitol, beside the Ara Coeli, at Rome. The view of that part of the castle which figures to-day as the back (it is the only aspect I had seen reproduced) exhibits the marks of restoration with the greatest assurance. The long facade, consisting only of balconied windows deeply recessed, erects itself on the summit of a considerable hill, which gives a fine, plunging movement to its foundations. The deep niches of the windows are all aglow with color. They have been repainted with red and blue, relieved with gold figures; and each of them looks more like the royal box at a theatre than like the aperture of a palace dark with memories. For all this, however, and in spite of the fact that, as in some others of the chateaux of Touraine, (always excepting the colossal Chambord, which is not in Touraine!) there is less vastness than one had expected, the least hospitable aspect of Blois is abundantly impressive. Here, as elsewhere, lightness and grace are the keynote; and the recesses of the windows, with their happy proportions, their sculpture, and their color, are the empty frames of brilliant pictures. They need the figure of a Francis I. to complete them, or of a Diane de Poitiers, or even of a Henry III. The base of this exquisite structure emerges from a bed of light verdure, which has been allowed to mass itself there, and which contributes to the springing look of the walls; while on the right it joins the most modern portion of the castle, the building erected, on foundations of enormous height and solidity, in 1635, by Gaston d'Orleans. This fine, frigid mansion the proper view of it is from the court within is one of the masterpieces of Francois Mansard, whom. a kind providence did not allow to make over the whole palace in the superior manner of his superior age. This had been a part of Gaston's plan, he was a blunderer born, and this precious project was worthy of him. This execution of it would surely have been one of the great misdeeds of history. Partially performed, the misdeed is not altogether to be regretted; for as one stands in the court of the castle, and lets one's eye wander from the splendid wing of Francis I. which is the last work of free and joyous invention to the ruled lines and blank spaces of the ponderous pavilion of Mansard, one makes one's reflections upon the advantage, in even the least personal of the arts, of having something to say, and upon the stupidity of a taste which had ended by becoming an aggregation of negatives. Gaston's wing, taken by itself, has much of the *bel air* which was to belong to the architecture of Louis XIV.; but, taken in contrast to its flowering, laughing, living neighbor, it marks the difference between inspiration and calculation. We scarcely grudge it its place, however, for it adds a price to the rest of the chateau.

We have entered the court, by the way, by jumping over the walls. The more orthodox method is to follow a modern, terrace, which leads to the left, from the side of the chateau that I began by speaking of, and passes round, ascending, to a little square on a considerably higher level, which is not, like a very modern square on which the back (as I have called it) looks out, a thoroughfare. This small, empty *place,* oblong in form, at once bright and quiet, with a certain grass-grown look, offers an excellent setting to the entrance-front of the palace, the wing of Louis XII. The restoration here has been lavish; but it was perhaps but an inevitable reaction against the injuries, still more lavish, by which the unfortunate building had long been overwhelmed. It had fallen into a state of ruinous neglect, relieved only by the misuse proceeding from successive generations of soldiers, for whom its charming chambers served as barrack-room. Whitewashed, mutilated, dishonored, the castle of Blois may be said to have escaped simply with its life. This is the history of Amboise as well, and is to a certain extent the history of Chambord. Delightful, at any rate, was the refreshed facade of Louis XII. as I stood and looked at it one bright September morning. In that soft, clear, merry light of Touraine, everything shows, everything speaks. Charming are the taste, the happy proportions, the color of this beautiful front, to which the new feeling for a purely domestic architecture an architecture of security and tranquility, in which art could indulge itself gave an air of youth and gladness. It is true that for a long time to come the castle of Blois was neither very safe nor very quiet; but its dangers came from within, from the evil passions of its inhabitants, and not from siege or invasion. The front of Louis XII. is of red brick, crossed here and there with purple; and the purple slate of the high roof, relieved with chimneys beautifully treated, and with the embroidered caps of pinnacles and arches, with the porcupine of Louis, the ermine and the festooned rope which formed the devices of Anne of Brittany, the tone of this rich-looking roof carries out the mild glow of the wall. The wide, fair windows look as if they had expanded to let in the rosy dawn of the Renaissance. Charming, for that matter, are the windows of all the chateaux of Touraine, with their squareness corrected (as it is not in the Tudor architecture) by the curve of the upper corners, which makes this line look above the expressive aperture like a penciled eyebrow. The low door of this front is crowned by a high, deep niche, in which, under a splendid canopy, stiffly astride of a stiffly draped charger, sits in profile an image of the good King Louis. Good as he had been, the father of his people, as he was called (I believe he remitted various taxes), he was not good enough to pass muster at the Revolution; and the effigy I have just described is no more than a reproduction of the primitive statue demolished at that period.

Pass beneath it into the court, and the sixteenth century closes round you. It is a pardonable flight of fancy to say that the expressive faces of an age in which human passions lay very near the surface seem to look out at you from the windows, from the balconies, from the thick foliage of the sculpture. The portion of the wing of Louis XII. that looks toward the court is supported on a deep arcade. On your right is the wing erected by Francis I., the reverse of the mass of building which you see on approaching the castle. This exquisite, this extravagant, this transcendent piece of architecture is the most joyous utterance of the French Renaissance. It is covered with an embroidery of sculpture, in which every detail is worthy of the hand of a goldsmith. In the middle of it, or rather a little to the left, rises the famous winding staircase (plausibly, but I believe not religiously, restored), which even the ages which most misused it must vaguely have admired. It forms a kind of chiseled cylinder, with wide interstices, so that the stairs are open to the air. Every inch of this structure, of its balconies, its pillars, its great central columns, is wrought over with lovely images, strange and ingenious devices, prime among which is the great heraldic salamander of Francis I. The salamander is everywhere at Blois, over the chimneys, over the doors, on the walls. This whole quarter , of the castle bears the stamp of that eminently pictorial prince. The running cornice along the top of the front is like all unfolded, an elongated, bracelet. The windows of the attic are like shrines for saints. The gargoyles, the medallions, the statuettes, the festoons, are like the elaboration of some precious cabinet rather than the details of a building exposed to the weather and to the ages. In the interior there is a profusion of restoration, and it is all restoration in color. This has been, evidently, a work of great energy and cost, but it will easily strike you as overdone. The universal freshness is a discord, a false note; it seems to light up the dusky past with an unnatural glare. Begun in the reign of Louis Philippe, this terrible process the more terrible always the more you admit that it has been necessary has been carried so far that there is now scarcely a square inch of the interior that has the color of the past upon it. It is true that the place had been so coated over with modern abuse that something was needed to keep it alive; it is only, perhaps, a pity that the restorers, not content with saving its life, should have undertaken to restore its youth. The love of consistency, in such a business, is a dangerous lure. All the old apartments have been rechristened, as it were; the geography of the castle has been re-established. The guardrooms, the bedrooms, the closets, the oratories, have recovered their identity. Every spot connected with the murder of the Duke of Guise is pointed out by a small, shrill boy, who takes you from room to room, and who has learned his lesson in perfection. The place is full of Catherine de' Medici, of Henry III., of memories, of ghosts, of echoes, of possible evocations and revivals. It is covered with crimson and gold. The fireplaces and the ceilings are magnificent; they look like expensive "sets" at the grand opera.

I should have mentioned that below, in the court, the front of the wing of Gaston d'Orleans faces you as you enter, so that the place is a course of French history. Inferior in beauty and grace to the other portions of the castle, the wing is yet a nobler monument than the memory of Gaston deserves. The second of the sons of Henry IV., who was no more fortunate as a father than as a husband, younger brother of Louis XIII., and father of the great Mademoiselle, the most celebrated, most ambitious, most self-complacent, and most unsuccessful *fille a marier* in French history, passed in enforced retirement at the castle of Blois the close of a life of clumsy intrigues against Cardinal Richelieu, in which his rashness was only equaled by his pusillanimity and his ill-luck by his inaccessibility to correction, and which, after so many follies and shames, was properly summed up in the project begun, but not completed of demolishing the beautiful habitation of his exile in order to erect a better one. With Gaston d'Orleans, however, who lived there without dignity, the history of the Chateau de Blois declines. Its interesting period is that of the wars of religion. It was the chief residence of Henry III., and the scene of the principal events of his depraved and dramatic reign. It has been restored more than enough, as I have said, by architects and decorators; the visitor, as he moves through its empty rooms, which are at once brilliant and ill-lighted (they have not been refurnished), undertakes a little restoration of his own. His imagination helps itself from the things that remain; he tries to see the life of the sixteenth century in its form and dress, its turbulence, its passions, its loves and hates, its treacheries, falsities, touches of faith, its latitude of personal development, its presentation of the whole nature, its nobleness of costume, charm of speech, splendor of taste, unequalled picturesqueness. The picture is full of movement, of contrasted light and darkness, full altogether of abominations. Mixed up with them all is the great name of religion, so that the drama wants nothing to make it complete. What episode was ever more perfect looked at as a dramatic occurrence than the murder of the Duke of Guise? The insolent prosperity of the victim; the weakness, the vices, the terrors, of the author of the deed; the perfect execution of the plot; the accumulation of horror in what followed it, give it, as a crime, a kind of immortal solidity.

But we must not take the Chateau de Blois too hard: I went there, after all, by way of entertainment. If among these sinister memories your visit should threaten to prove a tragedy, there is an excellent way of removing the impression. You may treat yourself at Blois to a very cheerful afterpiece. There is a charming industry practiced there, and practiced in charming conditions. Follow the bright little quay down the river till you get quite out of the town, and reach the point where the road beside the Loire becomes sinuous and attractive, turns the corner of diminutive headlands, and makes you wonder what is beyond. Let not your curiosity induce you, however, to pass by a modest white villa which overlooks the stream, enclosed in a fresh little court; for here dwells an artist, an artist in faience. There is no sort of sign, and the place looks peculiarly private. But if you ring at the gate, you will not be turned away. You will, on the contrary, be ushered upstairs into a parlor there is nothing resembling a shopencumbered with specimens of remarkably handsome pottery. The work is of the best, a careful reproduction of old forms, colors, devices; and the master of the establishment is one of those completely artistic types that are often found in France. His reception is as friendly as his work is ingenious; and I think it is not too much to say that you like the work the better because he has produced it. His vases, cups and jars, lamps, platters, *plaques,* with their brilliant glaze, their innumerable figures, their family likeness, and wide variations, are scattered, through his occupied rooms; they serve at once as his stock-in-trade and as household ornament. As we all know, this is an age of prose, of machinery, of wholesale production, of coarse and hasty processes. But one brings away from the establishment of the very intelligent M. Ulysse the sense of a less eager activity and a greater search for perfection. He has but a few workmen, and he gives them plenty of time. The place makes a little vignette, leaves an impression, the quiet white house in its garden on the road by the wide, clear river, without the smoke, the bustle, the ugliness, of so much of our modern industry. It ought to gratify Mr. Ruskin.

5

The second time I went to Blois I took a carriage for Chambord, and came back by the Chateau de Cheverny and the forest of Russy, a charming little expedition, to which the beauty of the afternoon (the finest in a rainy season that was spotted with bright days) contributed not a little. To go to Chambord, you cross the Loire, leave it on one side, and strike away through a country in which salient features become less and less numerous, and which at last has no other quality than a look of intense, and peculiar rurality, the characteristic, even when it is not the charm, of so much of the landscape of France. This is not the appearance of wildness, for it goes with great cultivation; it is simply the presence of the delving, drudging, economizing peasant. But it is a deep, unrelieved rusticity. It is a peasant's landscape; not, as in England, a landlord's. On the way to Chambord you enter the flat and sandy Sologne. The wide horizon opens out like a great *potager,* without interruptions, without an eminence, with here and there a long, low stretch of wood. There is an absence of hedges, fences, signs of property; everything is absorbed in the general flatness, the patches of vineyard, the scattered cottages, the villages, the children (planted and staring and almost always pretty), the women in the fields, the white caps, the faded blouses, the big sabots. At the end of an hour's drive (they assure you at Blois that even with two horses you will spend double that time), I passed through a sort of gap in a wall, which does duty as the gateway of the domain of an exiled pretender. I drove along a straight avenue, through a disfeatured park, the park of Chambord has twenty-one miles of circumference, a very sandy, scrubby, melancholy plantation, in which the timber must have been cut many times over and is to-day a mere tangle of brushwood. Here, as in so many spots in France, the traveler perceives that he is in a land of revolutions. Nevertheless, its great extent and the long perspective of its avenues give this desolate boscage a certain majesty; just as its shabbiness places it in agreement with one of the strongest impressions of the chateau. You follow one of these long perspectives a proportionate time, and at last you see the chimneys and pinnacles of Chambord rise apparently out of the ground. The filling-in of the wide moats that formerly surrounded it has, in vulgar parlance, let it down, bud given it an appearance of top heaviness that is at the same time a magnificent Orientalism. The towers, the turrets, the cupolas, the gables, the lanterns, the chimneys, look more like the spires of a city than the salient points of a single building. You emerge from the avenue and find yourself at the foot of an enormous fantastic mass. Chambord has a strange mixture of society and solitude. A little village clusters within view of its stately windows, and a couple of inns near by offer entertainment to pilgrims. These things, of course, are incidents of the political proscription which hangs its thick veil over the place. Chambord is truly royal, royal in its great scale, its grand air, its indifference to common considerations. If a cat may look at a king, a palace may lock at a tavern. I enjoyed my visit to this extraordinary structure as much as if I had been a legitimist; and indeed there is something interesting in any monument of a great system, any bold presentation of a tradition.

You leave your vehicle at one of the inns, which are very decent and tidy, and in which every one is very civil, as if in this latter respect the influence of the old regime pervaded the neighborhood, and you walk across the grass and the gravel to a small door, a door infinitely subordinate and conferring no title of any kind on those who enter it. Here you ring a bell, which a highly respectable person answers (a person perceptibly affiliated, again, to the old regime), after which she ushers you across a vestibule into an inner court. Perhaps the strongest impression I got at Chambord came to me as I stood in this court. The woman who admitted me did not come with me; I was to find my guide somewhere else. The specialty of Chambord is its prodigious round towers. There are, I believe, no less than eight of them, placed at each angle of the inner and outer square of buildings; for the castle is in the form of a larger structure which encloses a smaller one. One of these towers stood before me in the court; it seemed to fling its shadow over the place; while above, as I looked up, the pinnacles and gables, the enormous chimneys, soared into the bright blue air. The place was empty and silent; shadows of gargoyles, of extraordinary projections, were thrown across the clear gray surfaces. One felt that the whole thing was monstrous. A cicerone appeared, a languid young man in a rather shabby livery, and led me about with a mixture of the impatient and the desultory, of condescension and humility. I do not profess to understand the plan of Chambord, and I may add that I do not even desire to do so; for it is much more entertaining to think of it, as you can so easily, as an irresponsible, insoluble labyrinth. Within, it is a wilderness of empty chambers, a royal and romantic barrack. The exiled prince to whom it gives its title has not the means to keep up four hundred rooms; he contents himself with preserving the huge outside. The repairs of the prodigious roof alone must absorb a large part of his revenue. The great feature of the interior is the celebrated double staircase, rising straight through the building, with two courses of steps, so that people may ascend and descend without meeting. This staircase is a truly majestic piece of humor; it gives you the note, as it were, of Chambord. It opens on each landing to a vast guard-room, in four arms, radiations of the winding shaft. My guide made me climb to the great open-work lantern which, springing from the roof at the termination of the rotund staircase (surmounted here by a smaller one), forms the pinnacle of the bristling crown of Chambord. This lantern is tipped with a huge *fleur-de-lis* in stone, the only one, I believe, that the Revolution did not succeed in pulling down. Here, from narrow windows, you look over the wide, flat country and the tangled, melancholy park, with the rotation of its straight avenues. Then you walk about the roof, in a complication of galleries, terraces, balconies, through the multitude of chimneys and gables. This roof, which is in itself a sort of castle in the air, has an extravagant, fabulous quality, and with its profuse ornamentation, the salamander of Francis I. is a contant motive, its lonely pavements, its sunny niches, the balcony that looks down over the closed and grass-grown main entrance, a strange, half-sad, half-brilliant charm. The stone-work is covered with fine mould. There are places that reminded me of some of those quiet, mildewed corners of courts and terraces, into which the traveler who wanders through the Vatican looks down from neglected windows. They show you two or three furnished rooms, with Bourbon portraits, hideous tapestries from the ladies of France, a collection of the toys of the *enfant du miracle,* all military and of the finest make. "Tout cela fonctionne," the guide said of these miniature weapons; and I wondered, if he should take it into his head to fire off his little canon, how much harm the Comte de Chambord would do.

From below, the castle would look crushed by the redundancy of its upper protuberances if it were not for the enormous girth of its round towers, which appear to give it a robust lateral development. These towers, however, fine as they are in their way, struck me as a little stupid; they are the exaggeration of an exaggeration. In a building erected after the days of defense, and proclaiming its peaceful character from its hundred embroideries and cupolas, they seem to indicate a want of invention. I shall risk the accusation of bad taste if I say that, impressive as it is, the Chateau de Chambord seemed to me to have altogether a little of that quality of stupidity. The trouble is that it represents nothing very particular; it has not happened, in spite of sundry vicissitudes, to have a very interesting history. Compared with that of Blois and Amboise, its past is rather vacant; and one feels to a certain extent the contrast between its pompous appearance and its spacious but somewhat colorless annals. It had indeed the good fortune to be erected by Francis I., whose name by itself expresses a good deal of history. Why he should have built a palace in those sandy plains will ever remain an unanswered question, for kings have never been obliged to give reasons. In addition to the fact that the country was rich in game and that Francis was a passionate hunter, it is suggested by M. de la Saussaye, the author of the very complete little history of Chambord which you may buy at the bookseller's at Blois, that he was governed in his choice of the site by the accident of a charming woman having formerly lived there. The Comtesse de Thoury had a manor in the neighborhood, and the Comtesse de Thoury had been the object of a youthful passion on the part of the most susceptible of princes before his accession to the throne. This great pile was reared, therefore, according to M. de la Saussaye, as a *souvenir de premieres amours!* It is certainly a very massive memento; and if these tender passages were proportionate to the building that commemorates them, they were tender indeed. There has been much discussion as to the architect employed by Francis I., and the honor of having designed this splendid residence has been claimed for several of the Italian artists who early in the sixteenth century came to seek patronage in France. It seems well established to-day, however, that Chambord was the work neither of Primaticcio, of Vignola, nor of Il Rosso, all of whom have left some trace of their sojourn in France; but of an obscure yet very complete genius, Pierre Nepveu, known as Pierre Trinqueau, who is designated in the papers which preserve in some degree the history of the origin of the edifice, as the *maistre de l'oeuvre de maconnerie*. Behind this modest title, apparently, we must recognize one of the most original talents of the French Renaissance; and it is a proof of the vigor of the artistic life of that period that, brilliant production being everywhere abundant, an artist of so high a value should not have been treated by his contemporaries as a celebrity. We manage things very differently to-day.

The immediate successors of Francis I. continued to visit, Chambord; but it was neglected by Henry IV., and was never afterwards a favorite residence of any French king. Louis XIV. appeared there on several occasions, and the apparition was characteristically brilliant; but Chambord could not long detain a monarch who had gone to the expense of creating a Versailles ten miles from Paris. With Versailles, Fontainebleau, Saint-Germain, and Saint-Cloud within easy reach of their capital, the later French sovereigns had little reason to take the air in the dreariest province of their kingdom. Chambord therefore suffered from royal indifference, though in the last century a use was found for its deserted halls. In 1725 it was occupied by the luckless Stanislaus Leszczynski, who spent the greater part of his life in being elected King of Poland and being ousted from his throne, and who, at this time a refugee in France, had found a compensation for some of his misfortunes in marrying his daughter to Louis XV. He lived eight years at Chambord, and filled up the moats of the castle. In 1748 it found an illustrious tenant in the person of Maurice de Saxe, the victor of Fontenoy, who, however, two years after he had taken possession of it, terminated a life which would have been longer had he been less determined to make it agreeable. The Revolution, of course, was not kind to Chambord. It despoiled it in so far as possible of every vestige of its royal origin, and swept like a whirlwind through apartments to which upwards of two centuries had contributed a treasure of decoration and furniture. In that wild blast these precious things were destroyed or forever scattered. In 1791 an odd proposal was made to the French Government by a company of English Quakers who had conceived the bold idea of establishing in the palace a manufacture of some peaceful commodity not to-day recorded. Napoleon allotted Chambord, as a "dotation," to one of his marshals, Berthier, for whose benefit it was converted, in Napoleonic fashion, into the so-called principality of Wagram. By the Princess of Wagram, the marshal's widow, it was, after the Restoration, sold to the trustees of a national subscription which had been established for the purpose of presenting it to the infant Duke of Bordeaux, then prospective King of France. The presentation was duly made; but the

Comte de Chambord, who had changed his title in recognition of the gift, was despoiled of his property by the Government of Louis Philippe. He appealed for redress to the tribunals of his country; and the consequence of his appeal was an interminable litigation, by which, however, finally, after the lapse of twenty-five years, he was established in his rights. In 1871 he paid his first visit to the domain which had been offered him half a century before, a term of which he had spent forty years in exile. It was from Chambord that he dated his famous letter of the 5th of July of that year, the letter, directed to his socalled subjects, in which he waves aloft the white flag of the Bourbons. This amazing epistle, which is virtually an invitation to the French people to repudiate, as their national ensign, that immortal tricolor, the flag of the Revolution and the Empire, under which they have, won the glory which of all glories has hitherto been dearest to them, and which is associated with the most romantic, the most heroic, the epic, the consolatory, period of their history, this luckless manifesto, I say, appears to give the measure of the political wisdom of the excellent Henry V. It is the most factitious proposal ever addressed to an eminently ironical nation.

On the whole, Chambord makes a great impression; and the hour I was, there, while the yellow afternoon light slanted upon the September woods, there was a dignity in its desolation. It spoke, with a muffled but audible voice, of the vanished monarchy, which had been so strong, so splendid, but to-day has become a sort of fantastic vision, like the cupolas and chimneys that rose before me. I thought, while I lingered there, of all the fine things it takes to make up such a monarchy; and how one of them is a superfluity of moldering, empty, palaces. Chambord is touching, that is the best word for it; and if the hopes of another restoration are in the follies of the Republic, a little reflection on that eloquence of ruin ought to put the Republic on its guard. A sentimental tourist may venture to remark that in the presence of several chateaux which appeal in this mystical manner to the retrospective imagination, it cannot afford to be foolish. I thought of all this as I drove back to Blois by the way of the Chateau de Cheverny. The road took us out of the park of Chambord, but through a region of flat woodland, where the trees were not mighty, and again into the prosy plain of the Sologne, a thankless soil, all of it, I believe, but lately much amended by the magic of cheerful French industry and thrift. The light had already begun to fade, and my drive reminded me of a passage in some rural novel of Madame Sand. I passed a couple of timber and plaster churches, which looked very old, black, and crooked, and had lumpish wooden

27

porches and galleries encircling the base. By the time I reached Cheverny, the clear twilight had approached. It was late to ask to be allowed to visit an inhabited house; but it was the hour at which I like best to visit almost anything. My coachman drew up before a gateway, in a high wall, which opened upon a short avenue, along which I took my way on foot; the coachmen in those parts being, for reasons best known to themselves, mortally averse to driving up to a house. I answered the challenge of a very tidy little portress, who sat, in company with a couple of children, enjoying the evening air in, front of her lodge, and who told me to walk a little further and turn to the right. I obeyed her to the letter, and my turn brought me into sight of a house as charming as an old manor in a fairy tale. I had but a rapid and partial view of Cheverny; but that view was a glimpse of perfection. A light, sweet mansion stood looking over a wide green lawn, over banks of flowers and groups of trees. It had a striking character of elegance, produced partly by a series of Renaissance busts let into circular niches in the facade. The place looked so private, so reserved, that it seemed an act of violence to ring, a stranger and foreigner, at the graceful door. But if I had not rung I should be unable to express as it is such a pleasure to do my sense of the exceeding courtesy with which this admirable house is shown. It was near the dinner-hour, the most sacred hour of the day; but I was freely conducted into the inhabited apartments. They are extremely beautiful. What I chiefly remember is the charming staircase of white embroidered stone, and the great *salle des gardes* and *chambre a coucher du roi* on the second floor. Cheverny, built in 1634, is of a much later date than the other royal residences of this part of France; it belongs to the end of the Renaissance, and has a touch of the rococo. The guard-room is a superb apartment; and as it contains little save its magnificent ceiling and fireplace and certain dim tapestries on its walls, you the more easily take the measure of its noble proportions. The servant opened the shutters of a single window, and the last rays of the twilight slanted into the rich brown gloom. It was in the same picturesque fashion that I saw the bedroom (adjoining) of Henry IV., where a legendary-looking bed, draped in folds long unaltered, defined itself in the haunted dusk. Cheverny remains to me a very charming, a partly mysterious vision. I drove back to Blois in the dark, some nine miles, through the forest of Russy, which belongs to the State, and which, though consisting apparently of small timber, looked under the stars sufficiently vast and primeval. There was a damp autumnal smell and the occasional sound of a stirring thing; and as I moved through the evening air I thought of Francis I. and Henry IV.

6

You may go to Amboise either from Blois or from Tours; it is about half-way between these towns. The great point is to go, especially if you have put it off repeatedly; and to go, if possible, on a day when the great view of the Loire, which you enjoy from the battlements and terraces, presents itself under a friendly sky. Three persons, of whom the author of these lines was one, spent the greater part of a perfect Sunday morning in looking at it. It was astonishing, in the course of the rainiest season in the memory of the oldest Tourangeau, how many perfect days we found to our hand. The town of Amboise lies, like Tours, on the left bank of the river, a little whitefaced town, staring across an admirable bridge, and leaning, behind, as it were, against the pedestal of rock on which the dark castle masses itself. The town is so small, the pedestal so big, and the castle so high and striking, that the clustered houses at the base of the rock are like the crumbs that have fallen from a well-laden table. You pass among them, however, to ascend by a circuit to the chateau, which you attack, obliquely, from behind. It is the property of the Comte de Paris, another pretender to the French throne; having come to him remotely, by inheritance, from his ancestor, the Duc de Penthievre, who toward the close of the last century bought it from the crown, which had recovered it after a lapse. Like the castle of Blois it has been injured and defaced by base uses, but, unlike the castle of Blois, it has not been completely restored. "It is very, very dirty, but very curious," it is in these terms that I heard it described by an English lady, who was generally to be found engaged upon a tattered Tauchnitz in the little *salon de lecture* of the hotel at Tours. The description is not inaccurate; but it should be said that if part of the dirtiness of Amboise is the result of its having served for years as a barrack and as a prison, part of it comes from the presence of restoring stone-masons, who have woven over a considerable portion of it a mask of scaffolding. There is a good deal of neatness as well, and the restoration of some of the parts seems finished. This process, at Amboise, consists for the most part of simply removing the vulgar excrescences of the last two centuries.

The interior is virtually a blank, the old apartments having been chopped up into small modern rooms; it will have to be completely reconstructed. A worthy woman, with a military profile and that sharp, positive manner which the goodwives who show you through the chateaux of Touraine are rather apt to have, and in whose high respectability, to say nothing of the frill of her cap and the cut of her thick brown dress, my companions and I thought we discovered the particular note, or *nuance*, of Orleanism, a competent, appreciative, peremptory person, I say, attended us through the particularly delightful hour we spent upon the ramparts of Amboise. Denuded and disfeatured within, and bristling without with bricklayers' ladders, the place was yet extraordinarily impressive and interesting. I should confess that we spent a great deal of time in looking at the view. Sweet was the view, and magnificent; we preferred it so much to certain portions of the interior, and to occasional effusions of historical information, that the old lady with the prove sometimes lost patience with us. We laid ourselves open to the charge of preferring it even to the little chapel of Saint Hubert, which stands on the edge of the great terrace, and has, over the portal, a wonderful sculpture of the miraculous hunt of that holy man. In the way of plastic art this elaborate scene is the gem of Amboise. It seemed to us that we had never been in a place where there are so many points of vantage to look down from. In the matter of position Amboise is certainly supreme among the old houses of the Loire; and I say this with a due recollection of the claims of Chaumont and of Loches, which latter, by the way (excuse the afterthought), is not on the Loire. The platforms, the bastions, the terraces, the high-perched windows and balconies, the hanging gardens and dizzy crenellations, of this complicated structure, keep you in perpetual intercourse with an immense horizon. The great feature of the-place is the obligatory round tower which occupies the northern end of it, and which has now been, completely restored. It is of astounding size, a fortress in itself, and contains, instead of a staircase, a wonderful inclined plane, so wide and gradual that a coach and four may be driven to the top. This colossal cylinder has to-day no visible use; but it corresponds, happily enough, with the great circle of the prospect. The gardens of Amboise, perched in the air, covering the irregular remnants of the platform on which the castle stands, and making up in picturesqueness what they lack in extent, constitute of come but a scanty domain. But bathed, as we found them, in the autumn sunshine, and doubly private from their aerial site, they offered irresistible opportunities for a stroll, interrupted, as one leaned against their low parapets, by long, contemplative pauses. I remember, in particular, a certain terrace, planted with clipped limes, upon which we looked down from the summit of the big tower. It seemed from that point to be absolutely necessary to one's happiness to go down and spend the rest of the morning there; it was an ideal place to walk to and fro and talk. Our venerable conductress, to whom our relation had gradually become more filial, permitted us to gratify this innocent wish, to the extent, that is, of taking a turn or two under the mossy *tilleuls*. At the end of this terrace is the low door, in a wall, against the top of which, in 1498, Charles VIII., according to an accepted tradition, knocked his head to such good purpose that he died. It was within the walls of Amboise that his widow, Anne of Brittany, already in mourning for three children, two of whom we have seen commemorated in sepulchral marble at Tours, spent the first violence of that grief which was presently dispelled by a union with her husband's cousin and successor, Louis XII. Amboise was a frequent resort of the French Court during the sixteenth century; it was here that the young Mary Stuart spent sundry hours of her first marriage. The wars of religion have left here the ineffaceable stain which they left wherever they passed. An imaginative visitor at Amboise to-day may fancy that the traces of blood are mixed with the red rust on the crossed iron bars of the grim-looking balcony, to which the heads of the Huguenots executed on the discovery of the conspiracy of La Renaudie are rumored to have been suspended. There was room on the stout balustrade an admirable piece of work for a ghastly array. The same rumor represents Catherine de' Medici and the young queen as watching from this balcony the *noyades* of the captured Huguenots in the Loire. The facts of history are bad enough; the fictions are, if possible, worse; but there is little doubt that the future Queen of Scots learnt the first lessons of life at a horrible school. If in subsequent years she was a prodigy of innocence and virtue, it was not the fault of her whilom mother-in-law, of her uncles of the house of Guise, or of the examples presented to her either at the windows of the castle of Amboise or in its more private recesses.

It was difficult to believe in these dark deeds, however, as we looked through the golden morning at the placidity of the far-shining Loire. The ultimate consequence of this spectacle was a desire to follow the river as far as the castle of Chaumont. It is true that the cruelties practiced of old at Amboise might have seemed less phantasmal to persons destined to suffer from a modern form of inhumanity. The mistress of the little inn at the base of the castle-rock it stands very pleasantly beside the river, and we had breakfasted there declared to us that the Chateau de Chaumont, which is often during the autumn closed to visitors, was at that particular moment standing so wide open to receive us that it was our duty to hire one of her carriages and drive thither with speed. This assurance was so satisfactory that we presently found ourselves seated in this wily woman's most commodious vehicle, and rolling, neither too fast nor too slow, along the margin of the Loire. The drive of about an hour, beneath constant clumps of chestnuts, was charming enough to have been taken for itself; and indeed, when we reached Chaumont, we saw that our reward was to be simply the usual reward of virtue, the consciousness of having attempted the right. The Chateau de Chaumont was inexorably closed; so we learned from a talkative lodge-keeper, who gave what grace she could to her refusal. This good woman's dilemma was almost touching; she wished to reconcile two impossibles. The castle was not to be visited, for the family of its master was staying there; and yet she was loath to turn away a party of which she was good enough to say that it had a *grand genre;* for, as she also remarked, she had her living to earn. She tried to arrange a compromise, one of the elements of which was that we should descend from our carriage and trudge up a hill which would bring us to a designated point, where, over the paling of the garden, we might obtain an oblique and surreptitious view of a small portion of the castle walls. This suggestion led us to inquire (of each other) to what degree of baseness it is allowed to an enlightened lover of the picturesque to resort, in order to catch a glimpse of a feudal chateau. One of our trio decided, characteristically, against any form of derogation; so she sat in the carriage and sketched some object that was public property, while her two companions, who were not so proud, trudged up a muddy ascent which formed a kind of back-stairs. It is perhaps no more than they deserved that they were disappointed. Chaumont is feudal, if you please; but the modern spirit is in possession. It forms a vast clean-scraped mass, with big round towers, ungarnished with a leaf of ivy or a patch of moss, surrounded by gardens of moderate extent (save where the muddy lane of which I speak passes near it), and looking rather like an enormously magnified villa. The great merit of Chaumont is its position, which almost exactly resembles that of Amboise; it sweeps the river up and down, and seems to look over half the province. This, however, was better appreciated as, after coming down the hill and reentering the carriage, we drove across the long suspension-bridge which crosses the Loire just beyond the village, and over which we made our way to the small station of Onzain, at the farther end, to take the train back to Tours. Look back from the middle of this bridge; the whole picture composes, as the painters say. The towers, the pinnacles, the fair front of the chateau, perched above its fringe of garden and the rusty roofs of the village, and facing the afternoon sky, which is reflected also in the great stream that sweeps below, all this makes a contribution to your happiest memories of Touraine.

7

We never went to Chinon; it was a fatality. We planned it a dozen times; but the weather interfered, or the trains didn't suit, or one of the party was fatigued with the adventures of' the day before. This excursion was so much postponed that it was finally postponed to everything. Besides, we had to go to Chenonceaux, to Azay-le-Rideau, to Langeais, to Loches. So I have not the memory of Chinon; I have only the regret. But regret, as well as memory, has its visions; especially when, like memory, it is assisted by photographs. The castle of Chinon in this form appears to me as an enormous ruin, a mediaeval fortress, of the extent almost of a city. It covers a hill above the Vienne, and after being impregnable in its time is indestructible to-day. (I risk this phrase in the face of the prosaic truth. Chinon, in the days when it was a prize, more than once suffered capture, and at present it is crumbling inch by inch. It is apparent, however, I believe, that these inches encroach little upon acres of masonry.) It was in the castle that Jeanne Darc had her first interview with Charles VII., and it is in the town that Francois Rabelais is supposed to have been born. To the castle, moreover, the lover of the picturesque is earnestly recommended to direct his steps. But one cannot do everything, and I would rather have missed Chinon than Chenonceaux. Fortunate exceedingly were the few hours that we passed at this exquisite residence.

"In 1747," says Jean-Jacques Rousseau, in his "Confessions," "we went to spend the autumn in Touraine, at the Chateau, of Chenonceaux, a royal residence upon the Cher, built by Henry II. for Diana of Poitiers, whose initials are still to be seen there, and now in possession of M. Dupin, the farmer-general. We amused ourselves greatly in this fine spot; the living was of the best, and I became as fat as a monk. We made a great deal of music, and acted comedies."

This is the only description that Rousseau gives of one of the most romantic houses in France, and of an episode that must have counted as one of the most agreeable in his uncomfortable career. The eighteenth century contented itself with general epithets; and when Jean-Jacques has said that Chenonceaux was a "beau lieu," he thinks himself absolved from further characterization. We later sons of time have, both for our pleasure and our pain, invented the fashion of special terms, and I am afraid that even common decency obliges me to pay some larger tribute than this to the architectural gem of Touraine. Fortunately I can discharge my debt with gratitude. In going from Tours you leave the valley of the Loire and enter that of the Cher, and at the end of about an hour you see the turrets of the castle on your right, among the trees, down in the meadows, beside the quiet little river. The station and the village are about ten minutes' walk from the chateau, and the village contains a very tidy inn, where, if you are not in too great a hurry to commune with the shades of the royal favorite and the jealous queen, you will perhaps stop and order a dinner to be ready for you in the evening. A straight, tall avenue leads to the grounds of the castle; what I owe to exactitude compels me to add that it is crossed by the railway-line. The place is so arranged, however, that the chateau need know nothing of passing trains, which pass, indeed, though the grounds are not large, at a very sufficient distance. I may add that the trains throughout this part of France have a noiseless, desultory, dawdling, almost stationary quality, which makes them less of an offence than usual. It was a Sunday afternoon, and the light was yellow, save under the trees of the avenue, where, in spite of the waning of September, it was duskily green. Three or four peasants, in festal attire, were strolling about. On a bench at the beginning of the avenue, sat a man with two women. As I advanced with my companions he rose, after a sudden stare, and approached me with a smile, in which (to be Johnsonian for a moment) certitude was mitigated by modesty and eagerness was embellished with respect. He came toward me with a salutation that I had seen before, and I am happy to say that after an instant I ceased to be guilty of the brutality of not knowing where. There was only one place in the world where people smile like that, only one place where the art of salutation has that perfect grace. This excellent creature used to crook his arm, in Venice, when I stepped into my gondola; and I now laid my hand on that member with the familiarity of glad recognition; for it was only surprise that had kept me even for a moment from accepting the genial Francesco as an ornament of the landscape of Touraine. What on earth the phrase is the right one was a Venetian gondolier doing at Chenonceaux? He had been brought from Venice, gondola and all, by the mistress of the charming house, to paddle about on the Cher. Our meeting was affectionate, though there was a kind of violence in seeing him so far from home. He was too well dressed, too well fed; he had grown stout, and his nose had the tinge of good claret. He remarked that the life of the household to which he had the honor to belong was that of a *casa regia;* which must have been a great change for poor Checco, whose habits in Venice were not regal. However, he was the sympathetic Checco still; and for five minutes after I left him I thought less about the little pleasure-house by the Cher than about the palaces of the Adriatic.

But attention was not long in coming round to the charming structure that presently rose before us. The pale yellow front of the chateau, the small scale of which is at first a surprise, rises beyond a considerable court, at the entrance of which a massive and detached round tower, with a turret on its brow (a relic of the building that preceded the actual villa), appears to keep guard. This court is not enclosed or is enclosed, at least, only by the gardens, portions of which are at present in a state of violent reformation. Therefore, though Chenonceaux has no great height, its delicate facade stands up boldly enough. This facade, one of the most finished things in Touraine, consists of two stories, surmounted by an attic which, as so often in the buildings of the French Renaissance, is the richest part of the house. The high-pitched roof contains three windows of beautiful design, covered with embroidered caps and flowering into crocketed spires. The window above the door is deeply niched; it opens upon a balcony made in the form of a double pulpit, one of the most charming features of the front. Chenonceaux is not large, as I say, but into its delicate compass is packed a great deal of history, history which differs from that of Amboise and Blois in being of the private and sentimental kind. The echoes of the place, faint and far as they are to-day, are not political, but personal. Chenonceaux dates, as a residence, from the year 1515, when the shrewd Thomas Bohier, a public functionary who had grown rich in handling the finances of Normandy, and had acquired the estate from a family which, after giving it many feudal lords, had fallen into poverty, erected the present structure on the foundations of an old mill. The design is attributed, with I know not what justice, to Pierre Nepveu, *alias* Trinqueau, the audacious architect of Chambord. On the death of Bohier the house passed to his son, who, however, was forced, under cruel pressure, to surrender it to the crown, in compensation for a so-called deficit in the accounts of the late superintendent of the treasury. Francis I. held the place till his death; but Henry II., on ascending the throne, presented it out of hand to that mature charmer, the admired of two generations, Diana of Poitiers. Diana enjoyed it till the death of her protector; but when this event occurred, the widow of the monarch, who had been obliged to submit in silence, for years, to the ascendency of a rival, took the most pardonable of all the revenges with which the name of Catherine de' Medici is associated, and turned her out-of-doors. Diana was not in want of refuges, and Catherine went through the form of giving her Chaumont in exchange; but there was only one Chenonceaux. Catherine devoted herself to making the place more completely unique. The feature that renders it sole of its kind is not appreciated till you wander round to either side of the house. If a certain springing lightness is the characteristic of Chenonceaux, if it bears in every line the aspect of a place of recreation, a place intended for delicate, chosen pleasures, nothing can confirm this expression better than the strange, unexpected movement with which, from behind, it carries itself across the river. The earlier building stands in the water; it had inherited the foundations of the mill destroyed by Thomas Bohier. The first step, therefore, had been taken upon solid piles of masonry; and the ingenious Catherine she was a *raffinee* simply proceeded to take the others. She continued the piles to the opposite bank of the Cher, and over them she threw a long, straight gallery of two stories. This part of the chateau, which looks simply like a house built upon a bridge and occupying its entire length, is of course the great curiosity of Chenonceaux. It forms on each floor a charming corridor, which, within, is illuminated from either side by the flickering river-light. The architecture of these galleries, seen from without, is less elegant than that of the main building, but the aspect of the whole thing is delightful. I have spoken of Chenonceaux as a "villa," using the word advisedly, for the place is neither a castle nor a palace. It is a very exceptional villa, but it has the villa quality, the look of being intended for life in common. This look is not at all contradicted by the wing across the Cher, which only suggests intimate pleasures, as the French say, walks in pairs, on rainy days; games and dances on autumn nights; together with as much as may be of moonlighted dialogue (or silence) in the course, of evenings more genial still, in the well-marked recesses of windows.

It is safe to say that such things took place there in the last century, during the kindly reign of Monsieur and Madame Dupin. This period presents itself as the happiest in the annals of Chenonceaux. I know not what festive train the great Diana may have led, and my imagination, I am afraid, is only feebly kindled by the records of the luxurious pastimes organized on the banks of the Cher by the terrible daughter of the Medici, whose appreciation of the good things of life was perfectly consistent with a failure to perceive why others should live to enjoy, them. The best society that ever assembled there was collected at Chenonceaux during the middle of the eighteenth century. This was surely, in France at least, the age of good society, the period when it was well for appreciative people to have been born. Such people should of course have belonged to the fortunate few, and not to the miserable many; for the prime condition of a society being good is that it be not too large. The sixty years that preceded the French Revolution were the golden age of fireside talk and of those pleasures which proceed from the presence of women in whom the social art is both instinctive and acquired. The women of that period were, above all, good company; the fact is attested by a thousand documents. Chenonceaux offered a perfect setting to free conversation; and infinite joyous discourse must have mingled with the liquid murmur of the Cher. Claude Dupin was not only a great man of business, but a man of honor and a patron of knowledge; and his wife was gracious, clever, and wise. They had acquired this famous property by purchase (from one of the Bourbons; for Chenonceaux, for two centuries after the death of Catherine de' Medici, remained constantly in princely hands), and it was transmitted to their son, Dupin de Francueil, grandfather of Madame George Sand. This lady, in her Correspondence, lately published, describes a visit that she paid, more than thirty years ago, to those members of her family who were still in possession. The owner of Chenonceaux to-day is the daughter of an Englishman naturalized in France. But I have wandered far from my story, which is simply a sketch of the surface of the place. Seen obliquely, from either side, in combination with its bridge and gallery, the chateau is singular and fantastic, a striking example of a willful and capricious conception. Unfortunately, all caprices are not so graceful and successful, and I grudge the honor of this one to the false and blood-polluted Catherine. (To be exact, I believe the arches of the bridge were laid by the elderly Diana. It was Catherine, however, who completed the monument.) Within, the house has been, as usual, restored. The staircases and ceilings, in all the old royal residences of this part of France, are the parts that have suffered least; many of them have still much of the life of the old time about them. Some of the chambers of Chenonceaux, however, encumbered as they are with modern detail, derive a sufficiently haunted and suggestive look from the deep setting of their beautiful windows, which thickens the shadows and makes dark, corners. There is a charming little Gothic chapel, with its apse hanging over the water, fastened to the left flank of the house. Some of the upper balconies, which look along the outer face of the gallery, and either up or down the river, are delightful protected nooks. We walked through the lower gallery to the other bank of the Cher; this fine apartment appeared to be for the moment a purgatory of ancient furniture. It terminates rather abruptly; it simply stops, with a blank wall. There ought, of course, to have been a pavilion here, though I prefer very much the old defect to any modern remedy. The wall is not so blank, however, but that it contains a door which opens on a rusty drawbridge. This drawbridge traverses the small gap which divides the end of the gallery from the bank of the stream. The house, therefore, does not literally rest on opposite edges of the Cher, but rests on one and just fails to rest on the other. The pavilion would have made that up; but after a moment we ceased to miss this imaginary feature. We passed the little drawbridge, and wandered awhile beside the river. From this opposite bank the mass of the chateau looked more charming than ever; and the little peaceful, lazy Cher, where two or three men were fishing in the eventide, flowed under the clear arches and between the solid pedestals of the part that spanned it, with the softest, vaguest light on its bosom. This was the right perspective; we were looking across the river of time. The whole scene was deliciously mild. The moon came up; we passed back through the gallery and strolled about a little longer in the gardens. It was very still. I met my old gondolier in the twilight. He showed me his gondola; but I hated, somehow, to see it there. I don't like, as the French say, to *meler les genres*. A gondola in a little flat French river? The image was not less irritating, if less injurious, than the spectacle of a steamer in the Grand Canal, which had driven me away from Venice a year and a half before. We took our way back to the Grand Monarque, and waited in the little inn-parlor for a late train to Tours. We were not impatient, for we had an excellent dinner to occupy us; and even after we had dined we were still content to sit awhile and exchange remarks upon, the superior civilization of France. Where else, at a village inn, should we have fared so well? Where else should we have sat down to our refreshment without condescension? There were two or three countries in which it would not have been happy for us to arrive hungry, on a Sunday evening, at so modest an hostelry. At the little inn at Chenonceaux the *cuisine* was not only excellent, but the service was graceful. We were waited on by mademoiselle and her mamma; it was so that mademoiselle alluded

38

to the elder lady, as she uncorked for us a bottle of Vouvray mousseux. We were very comfortable, very genial; we even went so far as to say to each other that Vouvray mousseux was a delightful wine. From this opinion, indeed, one of our trio differed; but this member of the party had already exposed herself to the charge of being too fastidious, by declining to descend from the carriage at Chaumont and take that back-stairs view of the castle.

8

Without fastidiousness, it was fair to declare, on the other hand, that the little inn at Azay-le-Rideau was very bad. It was terribly dirty, and it was in charge of a fat *megere* whom the appearance of four trustful travelers we were four, with an illustrious fourth, on that occasion roused apparently to fury. I attached great importance to this incongruous hostess, for she uttered the only uncivil words I heard spoken (in connection with any business of my own) during a tour of some six weeks in France. Breakfast not at Azay-le-Rideau, therefore, too trustful traveler; or if you do so, be either very meek or very bold. Breakfast not, save under stress of circumstance; but let no circumstance whatever prevent you from going to see the admirable chateau, which is almost a rival of Chenonceaux. The village lies close to the gates, though after you pass these gates you leave it well behind. A little avenue, as at Chenonceaux, leads to the house, making a pretty vista as you approach the sculptured doorway. Azay is a most perfect and beautiful thing; I should place it third in any list of the great houses of this part of France in which these houses should be ranked according to charm. For beauty of detail it comes after Blois and Chenonceaux; but it comes before Amboise and Chambord. On the other hand, of course, it is inferior in majesty to either of these vast structures. Like Chenonceaux, it is a watery place, though it is more meagerly moated than the little chateau on the Cher. It consists of a large square *corps de logis*, with a round tower at each angle, rising out of a somewhat too slumberous pond. The water – the water of the Indre - surrounds it, but it is only on one side that it bathes its feet in the moat. On one of the others there is a little terrace, treated as a garden, and in front there is a wide court, formed by a wing which, on the right, comes forward. This front, covered with sculptures, is of the richest, stateliest effect. The court is approached by a bridge over the pond, and the house would reflect itself in this wealth of water if the water were a trifle less opaque. But there is a certain stagnation it affects more senses than one about the picturesque pools of Azay. On the hither side of the bridge is a garden, overshadowed by fine old sycamores, a garden shut in by greenhouses and by a fine last-century gateway, flanked with twin lodges. Beyond the chateau and the standing waters behind it is a so-called *parc*, which, however, it must be confessed, has little of park-like beauty. The old houses (many of them, that is) remain in France; but the old timber does not remain, and the denuded aspect of the few acres that surround the chateaux of Touraine is pitiful to the traveler who has learned to take the measure of such things from the manors and castles of England. The domain of the lordly Chaumont is that of an English suburban villa; and in that and in other places there is little suggestion, in the untended aspect of walk and lawns, of the vigilant British gardener. The manor of Azay, as seen to-day, dates from the early part of the sixteenth century; and the industrious Abbe Chevalier, in his very entertaining though slightly rose-colored book on Touraine,* (* Promenades pittoresque en Touraine. Tours: 1869.) speaks of it as, "perhaps the purest expression of the *belle Renaissance francaise.*" "Its height," he goes on, "is divided between two stories, terminating under the roof in a projecting entablature which imitates a row of machicolations. Carven chimneys and tall dormer windows, covered with imagery, rise from the roofs; turrets on brackets, of elegant shape, hang with the greatest lightness from the angles of the building. The soberness of the main lines, the harmony of the empty spaces and those that are filled out, the prominence of the crowning parts, the delicacy of all the details, constitute an enchanting whole." And then the Abbe speaks of the admirable staircase which adorns the north front, and which, with its extension, inside, constitutes the principal treasure of Azay. The staircase passes beneath one of the richest of porticos, a portico over which a monumental salamander indulges in the most decorative contortions. The sculptured vaults of stone which cover the windings of the staircase within, the fruits, flowers, ciphers, heraldic signs, are of the noblest effect. The interior of the chateau is rich, comfortable, extremely modern; but it makes no picture that compares with its external face, about which, with its charming proportions, its profuse yet not extravagant sculpture, there is something very tranquil and pure. I took particular fancy to the roof, high, steep, old, with its slope of bluish slate, and the way the weather-worn chimneys seemed to grow out of it, like living things out of a deep soil. The only defect of the house is the blankness and bareness of its walls, which have none of those delicate parasites attached to them that one likes to see on the surface of old dwellings. It is true that this bareness results in a kind of silvery whiteness of complexion, which carries out the tone of the quiet pools and even that of the scanty and shadeless park.

9

I hardly know what to say about the tone of Langeais, which, though I have left it to the end of my sketch, formed the objective point of the first excursion I made from Tours. Langeais is rather dark and gray; it is perhaps the simplest and most severe of all the castles of the Loire. I don't know why I should have gone to see it before any other, unless it be because I remembered the Duchesse de Langeais, who figures in several of Balzac's novels, and found this association very potent. The Duchesse de Langeais is a somewhat transparent fiction; but the castle from which Balzac borrowed the title of his heroine is an extremely solid fact. My doubt just above as to whether I should pronounce it exceptionally grey came from my having seen it under a sky which made most things look dark. I have, however, a very kindly memory of that moist and melancholy afternoon, which was much more autumnal than many of the days that followed it. Langeais lies down the Loire, near the river, on the opposite side from Tours, and to go to it you will spend half an hour in the train. You pass on the way the Chateau de Luynes, which, with its round towers catching the afternoon light, looks uncommonly well on a hill at a distance; you pass also the ruins of the castle of Cinq-Mars, the ancestral dwelling of the young favorite of Louis XIII., the victim, of Richelieu, the hero of Alfred de Vigny's novel, which is usually recommended to young ladies engaged in the study of French. Langeais is very imposing and decidedly somber; it marks the transition from the architecture of defense to that of elegance. It rises, massive and perpendicular, out of the centre of the village to which it gives its name, and which it entirely dominates; so that, as you stand before it, in the crooked and empty street, there is no resource for you but to stare up at its heavy overhanging cornice and at the huge towers surmounted with extinguishers of slate. If you follow this street to the end, however, you encounter in abundance the usual embellishments of a French village: little ponds or tanks, with women on their knees on the brink, pounding and thumping a lump of saturated linen; brown old crones, the tone of whose facial hide makes their nightcaps (worn by day) look dazzling; little alleys perforating the thickness of a row of cottages, and showing you behind, as a glimpse, the vividness of a green garden. In the rear of the castle rises a hill which must formerly have been occupied by some of its appurtenances, and which indeed is still partly enclosed within its court. You may walk round this eminence, which, with the small houses of the village at its base, shuts in the castle from behind. The enclosure is not defiantly guarded, however; for a small, rough path, which you presently reach, leads up to an open gate. This gate admits you to a vague and rather limited *parc*, which covers the crest of the hill, and through which you may walk into the gardens of castle. These gardens, of small extent, confront the dark walls with their brilliant parterres, and, covering the gradual slope of the hill, form, as it were, the fourth side of the court. This is the stateliest view of the chateau, which looks to you sufficiently grim and gray as, after asking leave of a neat young woman who sallies out to learn your errand, you sit there on a garden bench and take the measure of the three tall towers attached to this inner front and forming severally the cage of a staircase. The huge bracketed cornice (one of the features of Langeais) which is merely ornamental, as it is not machicolated, though it looks so, is continued on the inner face as well. The whole thing has a fine feudal air, though it was erected on the rains of feudalism.

The main event in the history of the castle is the marriage of Anne of Brittany to her first husband, Charles VIII., which took place in its great hall in 1491. Into this great hall we were introduced by the neat young woman, into this great hall and into sundry other halls, winding staircases, galleries, chambers. The cicerone of Langeais is in too great a hurry; the fact is pointed out in the excellent GuideJoanne. This ill-dissimulated vice, however, is to be observed, in the country of the Loire, in every one who carries a key. It is true that at Langeais there is no great occasion to indulge in the tourist's weakness of dawdling; for the apartments, though they contain many curious odds and ends of, antiquity, are not of first-rate interest. They are cold and musty, indeed, with that touching smell of old furniture, as all apartments should be through which the insatiate American wanders in the rear of a bored domestic, pausing to stare at a faded tapestry or to read the name on the frame of some simpering portrait.

To return to Tours my companion and I had counted on a train which (as is not uncommon in France) existed only in the "Indicateur des Chemins de Fer;" and instead of waiting for another we engaged a vehicle to take us home. A sorry *carriole* or *patache* it proved to be, with the accessories of a lumbering white mare and a little wizened, ancient peasant, who had put on, in honor of the occasion, a new blouse of extraordinary stiffness and blueness. We hired the trap of an energetic woman who put it "to" with her own hands; women in Touraine and the Blesois appearing to have the best of it in the business of letting vehicles, as well as in many other industries. There is, in fact, no branch of human activity in which one is not liable, in France, to find a woman engaged. Women, indeed, are not priests; but priests are, more or less; women. They are not in the army, it may be said; but then they *are* the army. They are very formidable. In France one must count with the women. The drive back from Langeais to Tours was long, slow, cold; we had an occasional spatter of rain. But the road passes most of the way close to the Loire, and there was something in our jog-trot through the darkening land, beside the flowing, river, which it was very possible to enjoy.

10

The consequence of my leaving to the last my little mention of Loches is that space and opportunity fail me; and yet a brief and hurried account of that extraordinary spot would after all be in best agreement with my visit. We snatched a fearful joy, my companion and I, the afternoon we took the train for Loches. The weather this time had been terribly against us: again and again a day that promised fair became hopelessly foul after lunch. At last we determined that if we could not make this excursion in the sunshine, we would make it with the aid of our umbrellas. We grasped them firmly and started for the station, where we were detained an unconscionable time by the evolutions, outside, of certain trains laden with liberated (and exhilarated) conscripts, who, their term of service ended, were about to be restored to civil life. The trains in Touraine are provoking; they serve as little as possible for excursions.

If they convey you one way at the right hour, it is on the condition of bringing you back at the wrong; they either allow you far too little time to examine the castle or the ruin, or they leave you planted in front of it for periods that outlast curiosity. They are perverse, capricious, exasperating. It was a question of our having but an hour or two at Loches, and we could ill afford to sacrifice to accidents. One of the accidents, however, was that the rain stopped before we got there, leaving behind it a moist mildness of temperature and a cool and lowering sky, which were in perfect agreement with the gray old city. Loches is certainly one of the greatest impressions of the traveler in central France, the largest cluster of curious things that presents itself to his sight. It rises above the valley of the Indre, the charming stream set in meadows and sedges, which wanders through the province of Berry and through many of the novels of Madame George Sand; lifting from the summit of a hill, which it covers to the base, a confusion of terraces, ramparts, towers, and spires. Having but little time, as I say, we scaled the hill amain, and wandered briskly through this labyrinth of antiquities. The rain had decidedly stopped, and save that we had our train on our minds, we saw Loches to the best advantage. We enjoyed that sensation with which the conscientious tourist is or ought to be well acquainted, and for which, at any rate, he has a formula in his rough-and-ready language. We "experienced," as they say, (most odious of verbs!) an "agreeable disappointment." We were surprised and delighted; we had not suspected that Loches was so good.

I hardly know what is best there: the strange and impressive little collegial church, with its Romanesque atrium or narthex, its doorways covered with primitive sculpture of the richest kind, its treasure of a so-called pagan altar, embossed with fighting warriors, its three pyramidal domes, so unexpected, so sinister, which I have not met elsewhere, in church architecture; or the huge square keep, of the eleventh century, the most cliff-like tower I remember, whose immeasurable thickness I did not penetrate; or the subterranean mysteries of two other less striking but not less historic dungeons, into which a terribly imperative little cicerone introduced us, with the aid of downward ladders, ropes, torches, warnings, extended hands; and, many, fearful anecdotes, all in impervious darkness. These horrible prisons of Loches, at an incredible distance below the daylight, were a favorite resource of Louis XI., and were for the most part, I believe, constructed by him. One of the towers of the castle is garnished with the hooks or supports of the celebrated iron cage in which he confined the Cardinal La Balue, who survived so much longer than might have been expected this extraordinary mixture of seclusion and exposure. All these things form part of the castle of Loches, whose enormous *enceinte* covers the whole of the top of the hill, and abounds in dismantled gateways, in crooked passages, in winding lanes that lead to postern doors, in long facades that look upon terraces interdicted to the visitor, who perceives with irritation that they command magnificent views. These views are the property of the sub-prefect of the department, who resides at the Chateau de Loches, and who has also the enjoyment of a garden a garden compressed and curtailed, as those of old castles that perch on hill-tops are apt to be containing a horse-chestnut tree of fabulous size, a tree of a circumference so vast and so perfect that the whole population of Loches might sit in concentric rows beneath its boughs. The gem of the place, however, is neither the big *marronier*, nor the collegial church, nor the mighty dungeon, nor the hideous prisons of Louis XI.; it is simply the tomb of Agnes Sorel, *la belle des belles*, so many years the mistress of Charles VII. She was buried, in 1450, in the collegial church, whence, in the beginning of the present century, her remains, with the monument that marks them, were transferred to one of the towers of the castle. She has always, I know not with what justice, enjoyed a fairer fame than most ladies who have occupied her position, and this fairness is expressed in the delicate statue that surmounts her tomb. It represents her lying there in lovely demureness, her hands folded with the best modesty, a little kneeling angel at either side of her head, and her feet, hidden in the folds of her decent robe, resting upon a pair of couchant lambs, innocent reminders of her name. Agnes, however, was not lamb-like, inasmuch as, according to popular tradition at least, she exerted herself sharply in favor of the expulsion of the English from France. It is one of the suggestions of Loches that the young Charles VII., hard put to it as he was for a treasury and a capital, "le roi de Bourges," he was called at Paris, was yet a rather privileged mortal, to stand up as he does before posterity between the noble Joan and the *gentille Agnes*, deriving, however much more honor from one of these companions than from the other. Almost as delicate a relic of antiquity as this fascinating tomb is the exquisite oratory of Anne of Brittany, among the apartments of the castle the only chamber worthy of note. This small room, hardly larger than a closet, and forming part of the addition made to the edifice by Charles VIII., is embroidered over with the curious and remarkably decorative device of the ermine and festooned cord. The objects in themselves are not especially graceful; but the constant repetition of the figure on the walls and ceiling produces an effect of richness, in spite of the modern whitewash with which, if I remember rightly, they have been endued. The little streets of Loches wander crookedly down the hill, and are full of charming pictorial "bits:" an old town gate, passing under a mediaeval tower, which is ornamented by Gothic windows and the empty niches of statues; a meager but delicate *hotel de ville*, of the Renaissance, nestling close beside it; a curious *chancellerie* of the middle of the sixteenth century, with mythological figures and a Latin inscription on the front, both of these latter buildings being rather unexpected features of the huddled and precipitous little town. Loches has a suburb on the other side of the Indre, which we had contented ourselves with looking down at from the heights, while we wondered whether, even if it had not been getting late and our train were more accommodating, we should care to take our way across the bridge and look up that bust, in terra-cotta, of Francis I., which is the principal ornament of the Chateau de Sansac and the faubourg of Beaulieu. I think we decided that we should not; that we were already quite well enough acquainted with the nasal profile of that monarch.

11

I know not whether the exact limits of an excursion, as distinguished from a journey, have ever been fixed; at any rate, it seemed none of my business, at Tours, to settle the question. Therefore, though the making of excursions had been the purpose of my stay, I thought it vain, while I started for Bourges, to determine to which category that little expedition might belong. It was not till the third day that I returned to Tours; and the distance, traversed for the most part after dark, was even greater than I had supposed. That, however, was partly the fault of a tiresome wait at Vierzon, where I had more than enough time to dine, very badly, at the *buffet*, and to observe the proceedings of a family who had entered my railway carriage at Tours and had conversed unreservedly, for my benefit, all the way from that station, a family whom it entertained me to assign to the class of *petite noblesse de province*. Their noble origin was confirmed by the way they all made *maigre* in the refreshment oom (it happened to be a Friday), as if it had been possible to do anything else. They ate two or three omelets apiece, and ever so many little cakes, while the positive, talkative mother watched her children as the waiter handed about the roast fowl. I was destined to share the secrets of this family to the end; for when I had taken place in the empty train that was in waiting to convey us to Bourges, the same vigilant woman pushed them all on top of me into my compartment, though the carriages on either side contained no travelers at all. It was better, I found, to have dined (even on omelets and little cakes) at the station at Vierzon than at the hotel at Bourges, which, when I reached it at nine o'clock at night, did not strike me as the prince of hotels. The inns in the smaller provincial towns in France are all, as the term is, commercial, and the *commis-voyageur* is in triumphant possession. I saw a great deal of him for several weeks after this; for he was apparently the only traveler in the southern provinces, and it was my daily fate to sit opposite to him at tables d'hote and in railway trains. He may be known by two infallible signs, his hands are fat, and he tucks his napkin into his shirt-collar. In spite of these idiosyncrasies, he seemed to me a reserved and inoffensive person, with singularly little of the demonstrative good-humor that he has been described as possessing. I saw no one who reminded me of Balzac's "illustre Gaudissart;" and indeed, in the course of a month's journey through a large part of France, I heard so little desultory conversation that I wondered whether a change had not come over the spirit of the people. They seemed to me as silent as Americans when Americans have not been "introduced," and infinitely less addicted to exchanging remarks in railway trains and at tables d'hote the colloquial and cursory English; a fact perhaps not worth mentioning were it not at variance with that reputation which the French have long enjoyed of being a pre-eminently sociable nation. The common report of the character of a people is, however, an indefinable product; and it is, apt to strike the traveler who observes for himself as very wide of the mark. The English, who have for ages been described (mainly by the French) as the dumb, stiff, unapproachable race, present to-day a remarkable appearance of good-humor and garrulity, and are distinguished by their facility of intercourse. On the other hand, any one who has seen half a dozen Frenchmen pass a whole day together in a railway carriage without breaking silence is forced to believe that the traditional reputation of these gentlemen is simply the survival of some primitive formula. It was true, doubtless, before the Revolution; but there have been great changes since then. The question of which is the better taste, to talk to strangers or to hold your tongue, is a matter apart; I incline to believe that the French reserve is the result of a more definite conception of social behavior. I allude to it only became it is at variance with the national fame, and at the same time is compatible with a very easy view of life in certain other directions. On some of these latter points the Boule d'Or at Bourges was full of instruction; boasting, as it did, of a hall of reception in which, amid old boots that had been brought to be cleaned, old linen that was being sorted for the wash, and lamps of evil odor that were awaiting replenishment, a strange, familiar, promiscuous household life went forward. Small scullions in white caps and aprons slept upon greasy benches; the Boots sat staring at you while you fumbled, helpless, in a row of pigeonholes, for your candlestick or your key; and, amid the coming and going of the *commis-voyageurs*, a little sempstress bent over the under-garments of the hostess, the latter being a heavy, stem, silent woman, who looked at people very hard.

It was not to be looked at in that manner that one had come all the way from Tours; so that within ten minutes after my arrival I sallied out into the darkness to get somehow and somewhere a happier impression. However late in the evening I may arrive at a place, I cannot go to bed without an impression. The natural place, at Bourges, to look for one seemed to be the cathedral; which, moreover, was the only thing that could account for my presence *dans cette galere*. I turned out of a small square, in front of the hotel, and walked up a narrow, sloping street, paved with big, rough stones and guiltless of a foot-way. It was a splendid starlight night; the stillness of a sleeping *ville de province* was over everything; I had the whole place to myself. I turned to my right, at the top of the street, where presently a short, vague lane brought me into sight of the cathedral. I approached it obliquely, from behind; it loomed up in the darkness above me, enormous and sublime. It stands on the top of the large but not lofty eminence over which Bourges is scattered, a very good position, as French cathedrals go, for they are not all so nobly situated as Chartres and Laon. On the side on which I approached it (the south) it is tolerably well exposed, though the precinct is shabby; in front, it is rather too much shut in. These defects, however, it makes up for on the north side and behind, where it presents itself in the most admirable manner to the garden of the Archeveche, which has been arranged as a public walk, with the usual formal alleys of the *jardin francais*. I must add that I appreciated these points only on the following day. As I stood there in the light of the stars, many of which had an autumnal sharpness, while others were shooting over the heavens, the huge, rugged vessel of the church overhung me in very much the same way as the black hull of a ship at sea would overhang a solitary swimmer. It seemed colossal, stupendous, a dark leviathan.

The next morning, which was lovely, I lost no time in going back to it, and found, with satisfaction, that the daylight did it no injury. The cathedral of Bourges is indeed magnificently huge; and if it is a good deal wanting in lightness and grace it is perhaps only the more imposing. I read in the excellent handbook of M. Joanne that it was projected "*des* 1172," but commenced only in the first years of the thirteenth century. "The nave" the writer adds, "was finished *tant bien que mal, faute de ressources;* the facade is of the thirteenth and fourteenth centuries in its lower part, and of the fourteenth in its upper." The allusion to the nave means the omission of the transepts. The west front consists of two vast but imperfect towers; one of which (the south) is immensely buttressed, so that its outline slopes forward, like that of a pyramid, being the taller of the two. If they had spires, these towers would be prodigious; as it is, given the rest of the church, they are wanting in elevation. There are five deeply recessed portals, all in a row, each surmounted with a gable; the gable over the central door being exceptionally high. Above the porches, which give the measure of its width, the front rears itself, piles itself, on a great scale, carried up by galleries, arches, windows, sculptures, and supported by the extraordinarily thick buttresses of which I have spoken, and which, though they embellish it with deep shadows thrown sidewise, do not improve its style. The portals, especially the middle one, are extremely interesting; they are covered with curious early sculptures. The middle one, however, I must describe alone. It has no less than six rows of figures, the others have four, some of which, notably the upper one, are still in their places. The arch at the top has three tiers of elaborate imagery. The upper of these is divided by the figure of Christ in judgment, of great size, stiff and terrible, with outstretched arms. On either side of him are ranged three or four angels, with the instruments of the Passion. Beneath him, in the second frieze, stands the angel of justice, with his scales; and on either side of him is the vision of the last judgment. The good prepare, with infinite titillation and complacency, to ascend to the skies; while the bad are dragged, pushed, hurled, stuffed, crammed, into pits and caldrons of fire. There is a charming detail in this section. Beside the angel, on, the right, where the wicked are the prey of demons, stands a little female figure, that of a child, who, with hands meekly folded and head gently raised, waits for the stern angel to decide upon her fate. In this fate, however, a dreadful, big devil also takes a keen interest; he seems on the point of appropriating the tender creature; he has a face like a goat and an enormous hooked nose. But the angel gently lays a hand upon the shoulder of the little girl the movement is full of dignity as if to say, "No; she belongs to the other side." The frieze below represents the general resurrection, with the good and the wicked emerging from their sepulchers. Nothing can be more quaint and charming than the difference shown in their way of responding to the final trump. The good get out of their tombs with a certain modest gayety, an alacrity tempered by respect; one of them kneels to pray as soon as he has disinterred himself. You may know the wicked, on the other hand, by their extreme shyness; they crawl out slowly and fearfully; they hang back, and seem to say, "Oh, dear!" These elaborate sculptures, full of ingenuous intention and of the reality of early faith, are in a remarkable state of preservation; they bear no superficial signs of restoration, and appear scarcely to have suffered from the centuries. They are delightfully expressive; the artist had the advantage of knowing exactly the effect he wished to produce.

The interior of the cathedral has a great simplicity and majesty, and, above all, a tremendous height. The nave is extraordinary in this respect; it dwarfs everything else I know. I should add, however, that I am, in architecture, always of the opinion of the last speaker. Any great building seems to me, while I look at it, the ultimate expression. At any rate, during the hour that I sat gazing along the high vista of Bourges, the interior of the great vessel corresponded to my vision of the evening before. There is a tranquil largeness, a kind of infinitude, about such an edifice: it soothes and purifies the spirit, it illuminates the mind. There are two aisles, on either side, in addition to the nave, five in all, and, as I have said, there are no transepts; an omission which lengthens the vista, so that from my place near the door the central jeweled window in the depths of the perpendicular choir seemed a mile or two away. The second, or outward, of each pair of aisles is too low, and the first too high; without this inequality the nave would appear to take an even more prodigious flight. The double aisles pass all the way round the choir, the windows of which are inordinately rich in magnificent old glass. I have seen glass as fine in other churches; but I think I have never seen so much of it at once.

Beside the cathedral, on the north, is a curious structure of the fourteenth or fifteenth century, which looks like an enormous flying buttress, with its support, sustaining the north tower. It makes a massive arch, high in the air, and produces a romantic effect as people pass under it to the open gardens of the Archeveche, which extend to a considerable distance in the rear of the church. The structure supporting the arch has the girth of a largeish house, and contains chambers with whose uses I am unacquainted, but to which the deep pulsations of the cathedral, the vibration of its mighty bells, and the roll of its organtones must be transmitted even through the great arm of stone.

The archiepiscopal palace, not walled in as at Tours, is visible as a stately habitation of the last century, now in course of reparation in consequence of a fire. From this side, and from the gardens of the palace, the nave of the cathedral is visible in all its great length and height, with its extraordinary multitude of supports. The gardens aforesaid, accessible through tall iron gates, are the promenade the Tuileries of the town, and, very pretty in themselves, are immensely set off by the overhanging church. It was warm and sunny; the benches were empty; I sat there a long time, in that pleasant state of mind which visits the traveler in foreign towns, when he is not too hurried, while he wonders where he had better go next. The straight, unbroken line of the roof of the cathedral was very noble; but I could see from this point how much finer the effect would have been if the towers, which had dropped almost out of sight, might have been carried still higher. The archiepiscopal gardens look down at one end over a sort of esplanade or suburban avenue lying on a lower level, on which they open, and where several detachments of soldiers (Bourges is full of soldiers) had just been drawn up. The civil population was also collecting, and I saw that something was going to happen. I learned that a private of the Chasseurs was to be "broken" for stealing, and every one was eager to behold the ceremony. Sundry other detachments arrived on the ground, besides many of the military who had come as a matter of taste. One of them described to me the process of degradation from the ranks, and I felt for a moment a hideous curiosity to see it, under the influence of which I lingered a little. But only a little; the hateful nature of the spectacle hurried me away, at the same time that others were hurrying forward. As I turned my back upon it I reflected that human beings are cruel brutes, though I could not flatter myself that the ferocity of the thing was exclusively French. In another country the concourse would have been equally great, and the moral of it all seemed to be that military penalties are as terrible as military honors are gratifying.

12

The cathedral is not the only lion of Bourges; the house of Jacques Coeur is an object of interest scarcely less positive. This remarkable man had a very strange history, and he too was "broken," like the wretched soldier whom I did not stay to see. He has been rehabilitated, however, by an age which does not fear the imputation of paradox, and a marble statue of him ornaments the street in front of his house. To interpret him according to this image a womanish figure in a long robe and a turban, with big bare arms and a dramatic pose would be to think of him as a kind of truculent sultana. He wore the dress of his period, but his spirit was very modern; he was a Vanderbilt or a Rothschild of the fifteenth century. He supplied the ungrateful Charles VII. with money to pay the troops who, under the heroic Maid, drove the English from French soil. His house, which to-day is used as a Palais de Justice, appears to have been regarded at the time it was built very much as the residence of Mr. Vanderbilt is regarded in New York to-day. It stands on the edge of the hill on which most of the town is planted, so that, behind, it plunges down to a lower level, and, if you approach it on that side, as I did, to come round to the front of it, you have to ascend a longish flight of steps. The back, of old, must have formed a portion of the city wall; at any rate, it offers to view two big towers, which Joanne says were formerly part of the defense of Bourges. From the lower level of which I speak the square in front of the post-office the palace of Jacques Coeur looks very big and strong and feudal; from the upper street, in front of it, it looks very handsome and delicate. To this street it presents two stories and a considerable length of facade; and it has, both within and without, a great deal of curious and beautiful detail. Above the portal, in the stonework, are two false windows, in which two figures, a man and a woman, apparently household servants, are represented, in sculpture, as looking down into the street. The effect is homely, yet grotesque, and the figures are sufficiently living to make one commiserate them for having been condemned, in so dull a town, to spend several centuries at the window. They appear to be watching for the return of their master, who left his beautiful house one morning and never came back.

The history of Jacques Coeur, which has been written by M. Pierre Clement, in a volume crowned by the French Academy, is very wonderful and interesting, but I have no space to go into it here. There is no more curious example, and few more tragical, of a great fortune crumbling from one day to the other, or of the antique superstition that the gods grow jealous of human success. Merchant, millionaire, banker, ship-owner, royal favorite, and minister of finance, explorer of the East and monopolist of the glittering trade between that quarter of the globe and his own, great capitalist who had anticipated the brilliant operations of the present time, he expiated his prosperity by poverty, imprisonment, and torture. The obscure points in his career have been elucidated by M. Clement, who has drawn, moreover, a very vivid picture of the corrupt and exhausted state of France during the middle of the fifteenth century. He has shown that the spoliation of the great merchant was a deliberately calculated act, and that the king sacrificed him without scruple or shame to the avidity of a singularly villainous set of courtiers. The whole story is an extraordinary picture of high-handed rapacity, the crudest possible assertion of the right of the stronger. The victim was stripped of his property, but escaped with his life, made his way out of France, and, betaking himself to Italy, offered his services to the Pope. It is proof of the consideration that he enjoyed in Europe, and of the variety of his accomplishments, that Calixtus III. should have appointed him to take command of a fleet which his Holiness was fitting out against the Turks. Jacques Coeur, however, was not destined to lead it to victory. He died shortly after the expedition had started, in the island of Chios, in 1456. The house of Bourges, his native place, testifies in some degree to his wealth and splendor, though it has in parts that want of space which is striking in many of the buildings of the Middle Ages. The court, indeed, is on a large scale, ornamented with turrets and arcades, with several beautiful windows, and with sculptures inserted in the walls, representing the various sources of the great fortune of the owner. M. Pierre Clement describes this part of the house as having been of an "incomparable richesse," an estimate of its charms which seems slightly exaggerated to-day. There is, however, something delicate and familiar in the bas-reliefs of which I have spoken, little scenes of agriculture and industry, which show, that the proprietor was not ashamed of calling attention to his harvests and enterprises. To-day we should question the taste of such allusions, even in plastic form, in the house of a "merchant prince" (say in the Fifth Avenue). Why is it, therefore, that these quaint little panels at Bourges do not displease us? It is perhaps because things very ancient never, for some mysterious reason, appear vulgar. This fifteenth-century millionaire, with his palace, his egotistical sculptures, may have produced that impression on some critical spirits of his own day.

The portress who showed me into the building was a dear little old woman, with the gentlest, sweetest, saddest face a little white, aged face, with dark, pretty eyes and the most considerate manner. She took me up into an upper hall, where there were a couple of curious chimney-pieces and a fine old oaken roof, the latter representing the hollow of a long boat. There is a certain oddity in a native of Bourges an inland town if there ever was one, without even a river (to call a river) to encourage nautical ambitions having found his end as admiral of a fleet; but this boat-shaped roof, which is extremely graceful and is repeated in another apartment, would suggest that the imagination of Jacques Coeur was fond of riding the waves. Indeed, as he trafficked in Oriental products and owned many galleons, it is probable that he was personally as much at home in certain Mediterranean ports as in the capital of the pastoral Berry. If, when he looked at the ceilings of his mansion, he saw his boats upside down, this was only a suggestion of the shortest way of emptying them of their treasures. He is presented in person above one of the great stone chimney-pieces, in company with his wife, Macee de Leodepart, I like to write such an extraordinary name. Carved in white stone, the two sit playing at chess at an open window, through which they appear to give their attention much more to the passers-by than to the game. They are also exhibited in other attitudes; though I do not recognize them in the composition on top of one of the fireplaces which represents the battlements of a castle, with the defenders (little figures between the crenellations) hurling down missiles with a great deal of fury and expression. It would have been hard to believe that the man who surrounded himself with these friendly and humorous devices had been guilty of such wrong-doing as to call down the heavy hand of justice.

It is a curious fact, however, that Bourges contains legal associations of a purer kind than the prosecution of Jacques Coeur, which, in spite of the rehabilitations of history, can hardly be said yet to have terminated, inasmuch as the law-courts of the city are installed in his quondam residence. At a short distance from it stands the Hotel Cujas, one of the curiosities of Bourges and the habitation for many years of the great jurisconsult who revived in the sixteenth century the study of the Roman law, and professed it during the close of his life in the university of the capital of Berry. The learned Cujas had, in spite of his sedentary pursuits, led a very wandering life; he died at Bourges in the year 1590. Sedentary pursuits is perhaps not exactly what I should call them, having read in the "Biographie Universelle" (sole source of my knowledge of the renowned Cujacius) that his usual manner of study was to spread himself on his belly on the floor. He did not sit down, he lay down; and the "Biographie Universelle" has (for so grave a work) an amusing picture of the short, fat, untidy scholar dragging himself *a plat ventre* across his room, from one pile of books to the other. The house in which these singular gymnastics took place, and which is now the headquarters of the gendarmerie, is one of the most picturesque at Bourges. Dilapidated and discolored, it has a charming Renaissance front. A high wall separates it from the street, and on this wall, which is divided by a large open gateway, are perched two overhanging turrets. The open gateway admits you to the court, beyond which the melancholy mansion erects itself, decorated also with turrets, with fine old windows, and with a beautiful tone of faded red brick and rusty stone. It is a charming encounter for a provincial bystreet; one of those accidents in the hope of which the traveler with a propensity for sketching (whether on a little paper block or on the tablets of his brain) decides to turn a corner at a venture. A brawny gendarme, in his shirt-sleeves, was polishing his boots in the court; an ancient, knotted vine, forlorn of its clusters, hung itself over a doorway, and dropped its shadow on the rough grain of the wall. The place was very sketchable. I am sorry to say, however, that it was almost the only "bit." Various other curious old houses are supposed to exist at Bourges, and I wandered vaguely about in search of them. But I had little success, and I ended by becoming skeptical. Bourges is a *ville de province* in the full force of the term, especially as applied invidiously. The streets, narrow, tortuous, and dirty, have very wide cobblestones; the houses for the most part are shabby, without local color. The look of things is neither modern nor antique, a kind of mediocrity of middle age. There is an enormous number of blank walls, walls of gardens, of courts, of private houses that avert themselves from the street, as if in natural chagrin at there being so little to see. Round about is a dull, flat, featureless country, on which the magnificent cathedral looks down. There is a peculiar dullness and ugliness in a French town of this type, which, I must immediately add, is not the most frequent one. In Italy, everything has a charm, a color, a grace; even desolation and *ennui*. In England a cathedral city may be sleepy, but it is pretty sure to be mellow. In the course of six weeks spent *en province*, however, I saw few places that had not more expression than Bourges.

I went back to the cathedral; that, after all, was a feature. Then I returned to my hotel, where it was time to dine, and sat down, as usual, with the *commisvoyageurs*, who cut their bread on their thumb and partook of every course; and after this repast I repaired for a while to the cafe, which occupied a part of the basement of the inn and opened into its court. This cafe was a friendly, homely, sociable spot, where it seemed the habit of the master of the establishment to *tutoyer* his customers, and the practice of the customers to *tutoyer* the waiter. Under these circumstances the waiter of course felt justified in sitting down at the same table with a gentleman who had come in and asked him for writing materials. He served this gentleman with a horrible little portfolio, covered with shiny black cloth and accompanied with two sheets of thin paper, three wafers, and one of those instruments of torture which pass in France for pens, these being the utensils invariably evoked by such a request; and then, finding himself at leisure, he placed himself opposite and began to write a letter of his own. This trifling incident reminded me afresh that France is a democratic country. I think I received an admonition to the same effect from the free, familiar way in which the game of whist was going on just behind me. It was attended with a great deal of noisy pleasantry, flavored every now and then with a dash of irritation. There was a young man of whom I made a note; he was such a beautiful specimen of his class. Sometimes he was very facetious, chattering, joking, punning, showing off; then, as the game went on and he lost, and had to pay the *consommation*, he dropped his amiability, slanged his partner, declared he wouldn't play any more, and went away in a fury. Nothing could be more perfect or more amusing than the contrast. The manner of the whole affair was such as, I apprehend, one would not have seen among our English-speaking people; both the jauntiness of the first phase and the petulance of the second. To hold the balance straight, however, I may remark that if the men were all fearful "cads," they were, with their cigarettes and their inconsistency, less heavy, less brutal, than our dear English-speaking cad; just as the bright little cafe where a robust materfamilias, doling out sugar and darning a stocking, sat in her place under the mirror behind the *comptoir*, was a much more civilized spot than a British publichouse, or a "commercial room," with pipes and whiskey, or even than an American saloon.

56

13

It is very certain that when I left Tours for Le Mans it was a journey and not an excursion; for I had no intention of coming back. The question, indeed, was to get away, no easy matter in France, in the early days of October, when the whole *jeunesse* of the country is going back to school. It is accompanied, apparently, with parents and grandparents, and it fills the trains with little pale-faced *lyceens*, who gaze out of the windows with a longing, lingering air, not unnatural on the part of small members of a race in which life is intense, who are about to be restored to those big educative barracks that do such violence to our American appreciation of the opportunities of boyhood. The train stopped every five minutes; but, fortunately, the country was charming, hilly and bosky, eminently good-humored, and dotted here and there with a smart little chateau. The old capital of the province of the Maine, which has given its name to a great American State, is a fairly interesting town, but I confess that I found in it less than I expected to admire. My expectations had doubtless been my own fault; there is no particular reason why Le Mans should fascinate. It stands upon a hill, indeed, a much better hill than the gentle swell of Bourges. This hill, however, is not steep in all directions; from the railway, as I arrived, it was not even perceptible. Since I am making comparisons, I may remark that, on the other hand, the Boule d'Or at Le Mans is an appreciably better inn than the Boule d'Or at Bourges. It looks out upon a small market-place which has a certain amount of character and seems to be slipping down the slope on which it lies, though it has in the middle an ugly *halle*, or circular markethouse, to keep it in position. At Le Mans, as at Bourges, my first business was with the cathedral, to which, I lost no time in directing my steps. It suffered by juxta-position to the great church I had seen a few days before; yet it has some noble features. It stands on the edge of the eminence of the town, which falls straight away on two sides of it, and makes a striking mass, bristling behind, as you see it from below, with rather small but singularly numerous flying buttresses. On my way to it I happened to walk through the one street which contains a few ancient and curious houses, a very crooked and untidy lane, of really mediaeval aspect, honored with the denomination of the Grand' Rue. Here is the house of Queen Berengaria, an absurd name, as the building is of a date some three hundred years later than the wife of Richard Coeur de Lion, who has a sepulchral monument in the south aisle of the cathedral. The structure in question very sketchable, if the sketcher could get far enough away from it is an elaborate little dusky facade, overhanging the street, ornamented with panels of stone, which are covered with delicate Renaissance sculpture. A fat old woman, standing in the door of a small grocer's shop next to it, a most gracious old woman, with a bristling moustache and a charming manner, told me what the house was, and also indicated to me a rotten-looking brown wooden mansion, in the same street, nearer the cathedral, as the Maison Scarron. The author of the "Roman Comique," and of a thousand facetious verses, enjoyed for some years, in the early part of his life, a benefice in the cathedral of Le Mans, which gave him a right to reside in one of the canonical houses. He was rather an odd canon, but his history is a combination of oddities. He wooed the comic muse from the arm-chair of a cripple, and in the same position he was unable even to go down on his knees prosecuted that other suit which made him the first husband of a lady of whom Louis XIV. was to be the second. There was little of comedy in the future Madame de Maintenon; though, after all, there was doubtless as much as there need have been in the wife of a poor man who was moved to compose for his tomb such an epitaph as this, which I quote from the "Biographie Universelle":-

"Celui qui cy maintenant dort, Fit plus de pitie que d'envie, Et souffrit mille fois la mort, Avant que de perdre la vie. Passant, ne fais icy de bruit, Et garde bien qu'il ne s'eveille, Car voicy la premiere nuit, Que le Pauvre Scarron sommeille."

There is rather a quiet, satisfactory *place* in front of the cathedral, with some good "bits" in it; notably a turret at the angle of one of the towers, and a very fine, steep-roofed dwelling, behind low walls, which it overlooks, with a tall iron gate. This house has two or three little pointed towers, a big, black, precipitous roof, and a general air of having had a history. There are houses which are scenes, and there are houses which are only houses. The trouble with the domestic architecture of the United States is that it is not scenic, thank Heaven! and the good fortune of an old structure like the turreted mansion on the hillside of Le Mans is that it is not simply a house. It is a person, as it were, as well. It would be well, indeed, if it might have communicated a little of its personality to the front of the cathedral, which has none of its own. Shabby, rusty, unfinished, this front has a Romanesque portal, but nothing in the way of a tower. One sees from without, at a glance, the peculiarity of the church, the disparity between the Romanesque nave, which is small and of the twelfth century, and the immense and splendid transepts and choir, of a period a hundred years later. Outside, this end of the church rises far above the nave, which looks merely like a long porch leading to it, with a small and curious Romanesque porch in its own south flank. The transepts, shallow but very lofty, display to the spectators in the *place* the reach of their two clere-story windows, which occupy, above, the whole expanse of the wall. The south transept terminates in a sort of tower, which is the only one of which the cathedral can boast. Within, the effect of the choir is superb; it is a church in itself, with the nave simply for a point of view. As I stood there, I read in my Murray that it has the stamp of the date of the perfection of pointed Gothic, and I found nothing to object to the remark. It suffers little by confrontation with Bourges, and, taken in itself, seems to me quite as fine. A passage of double aisles surrounds it, with the arches that divide them supported on very thick round columns, not clustered. There are twelve chapels in this passage, and a charming little lady chapel, filled with gorgeous old glass. The sustained height of this almost detached choir is very noble; its lightness and grace, its soaring symmetry, carry the eye up to places in the air from which it is slow to descend. Like Tours, like Chartres, like Bourges (apparently like all the French cathedrals, and unlike several English ones) Le Mans is rich in splendid glass. The beautiful upper windows of the choir make, far aloft, a sort of gallery of pictures, blooming with vivid color. It is the south transept that contains the formless image a clumsy stone woman lying on her back which purports to represent Queen Berengaria aforesaid.

The view of the cathedral from the rear is, as usual, very fine. A small garden behind it masks its base; but you descend the hill to a large *place de foire*, adjacent to a fine old pubic promenade which is known as Les Jacobins, a sort of miniature Tuileries, where I strolled for a while in rectangular alleys, destitute of herbage, and received a deeper impression of vanished things. The cathedral, on the pedestal of its hill, looks considerably farther than the fair-ground and the Jacobins, between the rather bare poles of whose straightly planted trees you may admire it at a convenient distance. I admired it till I thought I should remember it (better than the event has proved), and then I wandered away and looked at another curious old church, Notre-Dame-de-la-Couture. This sacred edifice made a picture for ten minutes, but the picture has faded now. I reconstruct a yellowish-brown facade, and a portal fretted with early sculptures; but the details have gone the way of all incomplete sensations. After you have stood awhile in the choir of the cathedral, there is no sensation at Le Mans that goes very far. For some reason not now to be traced, I had looked for more than this. I think the reason was to some extent simply in the name of the place; for names, on the whole, whether they be good reasons or not, are very active ones. Le Mans, if I am not mistaken, has a sturdy, feudal sound; suggests something dark and square, a vision of old ramparts and gates. Perhaps I had been unduly impressed by the fact, accidentally revealed to me, that Henry II., first of the English Plantagenets, was born there. Of course it is easy to assure one's self in advance, but does it not often happen that one had rather not be assured? There is a pleasure sometimes in running the risk of disappointment. I took mine, such as it was, quietly enough, while I sat before dinner at the door of one of the cafes in the market-place with a *bitter-et-curacao* (invaluable pretext at such an hour!) to keep me company. I remember that in this situation there came over me an impression which both included and excluded all possible disappointments. The afternoon was warm and still; the air was admirably soft. The good Manceaux, in little groups and pairs, were seated near me; my ear was soothed by the fine shades of French enunciation, by the detached syllables of that perfect tongue. There was nothing in particular in the prospect to charm; it was an average French view. Yet I felt a charm, a kind of sympathy, a sense of the completeness of French life and of the lightness and brightness of the social air, together with a desire to arrive at friendly judgments, to express a positive interest. I know not why this transcendental mood should have descended upon me then and there; but that idle half-hour in front of the cafe, in the mild October afternoon, suffused with human sounds, is perhaps the most definite thing I brought away from Le Mans.

14

I am shocked at finding, just after this noble declaration of principles that in a little note-book which at that time I carried about with me, the celebrated city of Angers is denominated a "sell." I reproduce this vulgar term with the greatest hesitation, and only because it brings me more quickly to my point. This point is that Angers belongs to the disagreeable class of old towns that have been, as the English say, "done up." Not the oldness, but the newness, of the place is what strikes the sentimental tourist to-day, as he wanders with irritation along second-rate boulevards, looking vaguely about him for absent gables. "Black Angers," in short, is a victim of modern improvements, and quite unworthy of its admirable name, a name which, like that of Le Mans, had always had, to my eyes, a highly picturesque value. It looks particularly well on the Shakespearean page (in "King John"), where we imagine it uttered (though such would not have been the utterance of the period) with a fine old insular accent. Angers figures with importance in early English history: it was the capital city of the Plantagenet race, home of that Geoffrey of Anjou who married, as second husband, the Empress Maud, daughter of Henry I. and competitor of Stephen, and became father of Henry II., first of the Plantagenet kings, born, as we have seen, at Le Mans. The facts create a natural presumption that Angers will look historic; I turned them over in my mind as I travelled in the train from Le Mans, through a country that was really pretty, and looked more like the usual English than like the usual French scenery, with its fields cut up by hedges and a considerable rotundity in its trees. On my way from the station to the hotel, however, it became plain that I should lack a good pretext for passing that night at the Cheval Blanc; I foresaw that I should have contented myself before the end of the day. I remained at the White Horse only long enough to discover that it was an exceptionally good provincial inn, one of the best that I encountered during six weeks spent in these establishments.

"Stupidly and vulgarly modernized," that is another phrase from my note-book, and note-books are not obliged to be reasonable. "There are some narrow and tortuous-streets, with a few curious old houses," I continue to quote; "there is a castle, of which the exterior is most extraordinary, and there is a cathedral of moderate interest. It is fair to say that the Chateau d'Angers is by itself worth a pilgrimage; the only drawback is that you have seen it in a quarter of an hour. You cannot do more than look at it, and one good look does your business. It has no beauty, no grace, no detail, nothing that charms or detains you; it is simply very old and very big, so big and so old that this simple impression is enough, and it takes its place in your recollections as a perfect specimen of a superannuated stronghold. It stands at one end of the town, surrounded by a huge, deep moat, which originally contained the waters of the Maine, now divided from it by a quay. The water-front of Angers is poor, wanting in color and in movement; and there is always an effect of perversity in a town lying near a great river and, yet not upon it. The Loire is a few miles off; but Angers contents itself with a meager affluent of that stream. The effect was naturally much better when the huge, dark mass of the castle, with its seventeen prodigious towers, rose out of the protecting flood. These towers are of tremendous girth and solidity; they are encircled with great bands, or hoops, of white stone, and are much enlarged at the base. Between them hang vast curtains of infinitely old-looking masonry, apparently a dense conglomeration of slate, the material of which the town was originally built (thanks to rich quarries in the neighborhood), and to which it owed its appellation of the Black. There are no windows, no apertures, and to-day no battlements nor roofs. These accessories were removed by Henry III., so that, in spite of its grimness and blackness, the place has not even the interest of looking like a prison; it being, as I supposed, the essence of a prison not to be open to the sky. The only features of the enormous structure are the black, somber stretches and protrusions of wall, the effect of which, on so large a scale, is strange and striking. Begun by Philip Augustus, and terminated by St. Louis, the Chateau d'Angers has of course a great deal of history. The luckless Fouquet, the extravagant minister of finance of Louis XIV., whose fall from the heights of grandeur was so sudden and complete, was confined here in 1661, just after his arrest, which had taken place at Nantes. Here, also, Huguenots and Vendeans have suffered effective captivity.

I walked round the parapet which protects the outer edge of the moat (it is all up hill, and the moat deepens and deepens), till I came to the entrance which faces the town, and which is as bare and strong as the rest. The concierge took me into the court; but there was nothing to see. The place is used as a magazine of ammunition, and the yard contains a multitude of ugly buildings. The only thing to do is to walk round the bastions for the view; but at the moment of my visit the weather was thick, and the bastions began and ended with themselves. So I came out and took another look at the big, black exterior, buttressed with white-ribbed towers, and perceived that a desperate sketcher might extract a picture from it, especially if he were to bring in, as they say, the little black bronze statue of the good King Rene (a weak production of David d'Angers), which, standing within sight, ornaments the melancholy faubourg. He would do much better, however, with the very striking old timbered house (I suppose of the fifteenth century) which is called the Maison d'Adam, and is easily the first specimen at Angers of the domestic architecture of the past. This admirable house, in the centre of the town, gabled, elaborately timbered, and much restored, is a really imposing monument. The basement is occupied by a linendraper, who flourishes under the auspicious sign of the Mere de Famille; and above his shop the tall front rises in five overhanging stories. As the house occupies the angle of a little *place*, this front is double, and the black beams and wooden supports, displayed over a large surface and carved and interlaced, have a high picturesqueness. The Maison d'Adam is quite in the grand style, and I am sorry to say I failed to learn what history attaches to its name. If I spoke just above of the cathedral as "moderate," I suppose I should beg its pardon; for this serious charge was probably prompted by the fact that it consists only of a nave, without side aisles. A little reflection now convinces me that such a form is a distinction; and, indeed, I find it mentioned, rather inconsistently, in my note-book, a little further on, as "extremely simple and grand." The nave is spoken of in the same volume as "big, serious, and Gothic," though the choir and transepts are noted as very shallow. But it is not denied that the air of the whole thing is original and striking; and it would therefore appear, after all, that the cathedral of Angers, built during the twelfth and thirteenth centuries, is a sufficiently honorable church; the more that its high west front, adorned with a very primitive Gothic portal, supports two elegant tapering spires, between which, unfortunately, an ugly modern pavilion has been inserted.

I remember nothing else at Angers but the curious old Cafe Serin, where, after I had had my dinner at the inn, I went and waited for the train which, at nine o'clock in the evening, was to convey me, in a couple of hours, to Nantes, an establishment remarkable for its great size and its air of tarnished splendor, its brown gilding and smoky frescos, as also for the fact that it was hidden away on the second floor of an unassuming house in an unilluminated street. It hardly seemed a place where you would drop in; but when once you had found it, it presented itself, with the cathedral, the castle, and the Maison d'Adam, as one of the historical monuments of Angers.

15

If I spent two nights at Nantes, it was for reasons of convenience rather than of sentiment; though, indeed, I spent them in a big circular room which had a stately, lofty, last-century look, a look that consoled me a little for the whole place being dirty. The high, old-fashioned, inn (it had a huge, windy *portecochere*, and you climbed a vast black stone staircase to get to your room) looked out on a dull square, surrounded with other tall houses, and occupied on one side by the theatre, a pompous building, decorated with columns and statues of the muses. Nantes belongs to the class of towns which are always spoken of as "fine," and its position near the mouth of the Loire gives it, I believe, much commercial movement. It is a spacious, rather regular city, looking, in the parts that I traversed, neither very fresh nor very venerable. It derives its principal character from the handsome quays on the Loire, which are overhung with tall eighteenth-century houses (very numerous, too, in the other streets), houses, with big *entresols* marked by arched windows, classic pediments, balcony rails of fine old iron-work. These features exist in still better form at Bordeaux; but, putting Bordeaux aside, Nantes is quite architectural. The view up and down the quays has the cool, neutral tone of color that one finds so often in French water-side places, the bright grayness which is the tone of French landscape art. The whole city has rather a grand, or at least an eminently well-established air. During a day passed in it of course I had time to go to the Musee; the more so that I have a weakness for provincial museums, a sentiment that depends but little on the quality of the collection. The pictures may be bad, but the place is often curious; and, indeed, from bad pictures, in certain moods of the mind, there is a degree of entertainment to be derived. If they are tolerably old they are often touching; but they must have a relative antiquity, for I confess I can do nothing with works of art of which the badness is of recent origin. The cool, still, empty chambers in which indifferent collections are apt to be preserved, the red brick tiles, the diffused light, the musty odor, the mementos around you of dead fashions, the snuffy custodian in a black skull cap, who pulls aside a faded curtain to show you the lusterless gem of the museum, these things have a mild historical quality, and the sallow canvases after all illustrate something. Many of those in the museum of Nantes illustrate the taste of a successful warrior; having been bequeathed to the city by

64

Napoleon's marshal, Clarke (created Duc de Feltre). In addition to these there is the usual number of specimens of the contemporary French school, culled from the annual Salons and presented to the museum by the State. Wherever the traveler goes, in France, he is reminded of this very honorable practice, the purchase by the Government of a certain number of "pictures of the year," which are presently distributed in the provinces. Governments succeed each other and bid for success by different devices; but the "patronage of art" is a plank, as we should say here, in every platform. The works of art are often ill-selected, there is an official taste which you immediately recognize, but the custom is essentially liberal, and a government which should neglect it would be felt to be painfully common. The only thing in this particular Musee that I remember is a fine portrait of a woman, by Ingres, very flat and Chinese, but with an interest of line and a great deal of style.

There is a castle at Nantes which resembles in some degree that of Angers, but has, without, much less of the impressiveness of great size, and, within, much more interest of detail. The court contains the remains of a very fine piece of late Gothic, a tall elegant building of the sixteenth century. The chateau is naturally not wanting in history. It was the residence of the old Dukes of Brittany, and was brought, with the rest of the province, by the Duchess Anne, the last representative of that race, as her dowry, to Charles VIII. I read in the excellent hand-book of M. Joanne that it has been visited by almost every one of the kings of France, from Louis XI. downward; and also that it has served as a place of sojourn less voluntary on the part of various other distinguished persons, from the horrible Merechal de Retz, who in the fifteenth century was executed at Nantes for the murder of a couple of hundred young children, sacrificed in abominable rites, to the ardent Duchess of Berry, mother of the Count of Chambord, who was confined there for a few hours in 1832, just after her arrest in a neighboring house. I looked at the house in question you may see it from the platform in front of the chateau and tried to figure to myself that embarrassing scene. The duchess, after having unsuccessfully raised the standard of revolt (for the exiled Bourbons), in the legitimist Bretagne, and being "wanted," as the phrase is, by the police of Louis Philippe, had hidden herself in a small but loyal house at Nantes, where, at the end of five months of seclusion, she was betrayed, for gold, to the austere M. Guizot, by one of her servants, an Alsatian Jew named Deutz. For many hours before her capture she had been compressed into an interstice behind a fireplace, and by the time she was drawn forth into the light she had been ominously scorched. The man who showed me the castle indicated also another historic spot, a house with little *tourelles*, on the Quai de la Fosse, in which Henry IV. is said to have signed the Edict of Nantes. I am, however, not in a position to answer for this pedigree.

There is another point in the history of the fine old houses which command the Loire, of which, I suppose, one may be tolerably sure; that is, their having, placid as they stand there to-day, looked down on the horrors of the Terror of 1793, the bloody reign of the monster Carrier and his infamous *noyades*. The most hideous episode of the Revolution was enacted at Nantes, where hundreds of men and women, tied together in couples, were set afloat upon rafts and sunk to the bottom of the Loire. The tall eighteenth-century house, full of the *air noble*, in France always reminds me of those dreadful years, of the street-scenes of the Revolution. Superficially, the association is incongruous, for nothing could be more formal and decorous than the patent expression of these eligible residences. But whenever I have a vision of prisoners bound on tumbrels that jolt slowly to the scaffold, of heads carried on pikes, of groups of heated *citoyennes* shaking their fists at closed coach-windows, I see in the background the well-ordered features of the architecture of the period, the clear gray stone, the high pilasters, the arching lines of the *entresol*, the classic pediment, the slate-covered attic. There is not much architecture at Nantes except the domestic. The cathedral, with a rough west front and stunted towers, makes no impression as you approach it. It is true that it does its best to recover its reputation as soon as you have passed the threshold. Begun in 1434 and finished about the end of the fifteenth century, as I discover in Murray, it has a magnificent nave, not of great length, but of extraordinary height and lightness. On the other hand, it has no choir whatever. There is much entertainment in France in seeing what a cathedral will take upon itself to possess or to lack; for it is only the smaller number that have the full complement of features. Some have a very fine nave and no choir; others a very fine choir and no nave. Some have a rich outside and nothing within; others a very blank face and a very glowing heart. There are a hundred possibilities of poverty and wealth, and they make the most unexpected combinations.

The great treasure of Nantes is the two noble sepulchral monuments which occupy either transept, and one of which has (in its nobleness) the rare distinction of being a production of our own time. On the south side stands the tomb of Francis II., the last of the Dukes of Brittany, and of his second wife, Margaret of Foix, erected in 1507 by their daughter Anne, whom we have encountered already at the Chateau de Nantes, where she was born; at Langeais, where she married her first husband; at Amboise, where she lost him; at Blois, where she married her second, the "good" Louis XII., who divorced an impeccable spouse to make room for her, and where she herself died. Transferred to the cathedral from a demolished convent, this monument, the masterpiece of Michel Colomb, author of the charming tomb of the children of Charles VIII. and the aforesaid Anne, which we admired at Saint Gatien of Tours, is one of the most brilliant works of the French Renaissance. It has a splendid effect, and is in perfect preservation. A great table of black marble supports the reclining figures of the duke and duchess, who lie there peacefully and majestically, in their robes and crowns, with their heads each on a cushion, the pair of which are supported, from behind, by three, charming little kneeling angels; at the foot of the quiet couple are a lion and a greyhound, with heraldic devices. At each of the angles of the table is a large figure in white marble of a woman elaborately dressed, with a symbolic meaning, and these figures, with their contemporary faces and clothes, which give them the air of realistic portraits, are truthful and living, if not remarkably beautiful. Round the sides of the tomb are small images of the apostles. There is a kind of masculine completeness in the work, and a certain robustness of taste.

In nothing were the sculptors of the Renaissance more fortunate than in being in advance of us with their tombs: they have left us noting to say in regard to the great final contrast, the contrast between the immobility of death and the trappings and honors that survive. They expressed in every way in which it was possible to express it the solemnity, of their conviction that the Marble image was a part of the personal greatness of the defunct, and the protection, the redemption, of his memory. A modern tomb, in comparison, is a skeptical affair; it insists too little on the honors. I say this in the face of the fact that one has only to step across the cathedral of Nantes to stand in the presence of one of the purest and most touching of modern tombs. Catholic Brittany has erected in the opposite transept a monument to one of the most devoted of her sons, General de Lamoriciere, the defender of the Pope, the vanquished of Castelfidardo. This noble work, from the hand of Paul Dubois, one of the most interesting of that new generation of sculptors who have revived in France an art of which our overdressed century had begun to despair, has every merit but the absence of a certain prime feeling. It is the echo of an earlier tune, an echo with a beautiful cadence. Under a Renaissance canopy of white marble, elaborately worked with arabesques and cherubs, in a relief so low that it gives the work a certain look of being softened and worn by time, lies the body of the Breton soldier, with, a crucifix clasped to his breast and a shroud thrown over his body. At each of the angles sits a figure in bronze, the two best of which, representing Charity and Military Courage, had given me extraordinary pleasure when they were exhibited (in the clay) in the Salon of 1876. They are admirably cast, and they have a certain greatness: the one, a serene, robust young mother, beautiful in line and attitude; the other, a lean and vigilant young man, in a helmet that overshadows his serious eyes, resting an outstretched arm, an admirable military member, upon the hilt of a sword. These figures contain abundant assurance that M. Paul Dubois has been attentive to Michael Angelo, whom we have all heard called a splendid example but a bad model. The visor-shadowed face of his warrior is more or less a reminiscence of the figure on the tomb of Lorenzo de' Medici at Florence; but it is doubtless none the worse for that. The interest of the work of Paul Dubois is its peculiar seriousness, a kind of moral good faith which is not the commonest feature of French art, and which, united as it is in this case with exceeding knowledge and a remarkable sense of form, produces an impression, of deep refinement. The whole monument is a proof of exquisitely careful study; but I am not sure that this impression on the part of the spectator is altogether a happy one. It explains much of its great beauty, and it also explains, perhaps, a little of a certain weakness. That word, however, is scarcely in place; I only mean that M. Dubois has made a visible effort, which has been most fruitful. Simplicity is not always strength, and our complicated modern genius contains treasures of intention. This fathomless modern element is an immense charm on the part of M. Paul Dubois. I am lost in admiration of the deep aesthetic experience, the enlightenment of taste, revealed by such work. After that, I only hope that Giuseppe Garibaldi may have a monument as fair.

16

To go from Nantes to La Rochelle you travel straight southward, across the historic *bocage* of La Vendee, the home of royalist bush-fighting. The country, which is exceedingly pretty, bristles with copses, orchards, hedges, and with trees more spreading and sturdy than the traveler is apt to deem the feathery foliage of France. It is true that as I proceeded it flattened out a good deal, so that for an hour there was a vast featureless plain, which offered me little entertainment beyond the general impression that I was approaching the Bay of Biscay (from which, in reality, I was yet far distant). As we drew near La Rochelle, however, the prospect brightened considerably, and the railway kept its course beside a charming little canal, or canalized river, bordered with trees, and with small, neat, bright-colored, and yet old-fashioned cottages and villas, which stood back on the further side, behind small gardens, hedges, painted palings, patches of turf. The whole effect was Dutch and delightful; and in being delightful, though not in being Dutch, it prepared me for the charms of La Rochelle, which from the moment I entered it I perceived to be a fascinating little town, a most original mixture of brightness and dullness. Part of its brightness comes from its being extraordinarily clean, in which, after all, it *is* Dutch; a virtue not particularly noticeable at Bourges, Le Mans, and Angers. Whenever I go southward, if it be only twenty miles, I begin to look out for the south, prepared as I am to find the careless grace of those latitudes even in things of which it may, be said that they may be south of something, but are not southern. To go from Boston to New York (in this state of mind) is almost as soft a sensation as descending the Italian side, of the Alps; and to go from New York to Philadelphia is to enter a zone of tropical luxuriance and warmth. Given this absurd disposition, I could not fail to flatter myself, on reaching La Rochelle, that I was already in the Midi, and to perceive in everything, in the language of the country, the *caractere meridional.* Really, a great many things had a hint of it. For that matter, it seems to me that to arrive in the south at a bound to wake up there, as it were would be a very imperfect pleasure. The full pleasure is to approach by stages and gradations; to observe the successive shades of difference by which it ceases to be the north. These shades are exceedingly fine, but your true south-lover has an eye for them all. If he perceive them at New York and Philadelphia, we imagine him boldly as liberated from Boston, how could he fail to perceive them at La Rochelle? The streets of this dear little city are lined with arcades, good, big, straddling arcades of stone, such as befit a land of hot summers, and which recalled to me, not to go further, the dusky portions of Bayonne. It contains, moreover, a great wide *place d'armes*, which looked for all the world like the piazza of some dead Italian town, empty, sunny, grass-grown, with a row of yellow houses overhanging it, an unfrequented cafe, with a striped awning, a tall, cold, florid, uninteresting cathedral of the eighteenth century on one side, and on the other a shady walk, which forms part of an old rampart. I followed this walk for some time, under the stunted trees, beside the grass-covered bastions; it is very charming, winding and wandering, always with trees. Beneath the rampart is a tidal river, and on the other side, for a long distance, the mossy walls of the immense garden of a seminary. Three hundred years ago, La Rochelle was the great French stronghold of Protestantism; but to-day it appears to be a'nursery of Papists.

The walk upon the rampart led me round to one of the gatesi of the town, where I found some small modern, fortifications and sundry red-legged soldiers, and, beyond the fortifications, another shady walk, a *mail*, as the French say, as well as a *champ de manoeuvre*, on which latter expanse the poor little red-legs were doing their exercise. It was all very quiet and very picturesque, rather in miniature; and at once very tidy and a little out of repair. This, however, was but a meager back-view of La Rochelle, or poor side-view at best. There are other gates than the small fortified aperture just mentioned; one of them, an old gray arch beneath a fine clock-tower, I had passed through on my way from the station. This picturesque Tour de l'Horloge separates the town proper from the port; for beyond the old gray arch, the place presents its bright, expressive little face to the sea. I had a charming walk about the harbor, and along the stone piers and sea-walls that shut it in. This indeed, to take things in their order, was after I had had my breakfast (which I took on arriving) and after I had been to the *hotel de ville*. The inn had a long narrow garden behind it, with some very tall trees; and passing through this garden to a dim and secluded *salle a manger*, buried in the heavy shade, I had, while I sat at my repast, a feeling of seclusion which amounted almost to a sense of incarceration. I lost this sense, however, after I had paid my bill, and went out to look for traces of the famous siege, which is the principal title of La Rochelle to renown. I had come thither partly because I thought it would be interesting to stand for a few moments in so gallant a spot, and partly because, I confess, I had a curiosity to see what had been the starting-point of the Huguenot emigrants who founded the town of New Rochelle in the State of New York, a place in which I had passed certain memorable hours. It was strange to think, as I strolled through the peaceful little port, that these quiet waters, during the wars of religion, had swelled with a formidable naval power. The Rochelais had fleets and admirals, and their stout little Protestant bottoms carried defiance up and down.

To say that I found any traces of the siege would be to misrepresent the taste for vivid whitewash by which La Rochelle is distinguished to-day. The only trace is the dent in the marble top of the table on which, in the *hotel de ville*, Jean Guiton, the mayor of the city, brought down his dagger with an oath, when in 1628 the vessels and regiments of Richelieu closed about it on sea and land. This terrible functionary was the soul of the resistance; he held out from February to October, in the midst of pestilence and famine. The whole episode has a brilliant place among the sieges of history; it has been related a hundred times, and I may only glance at it and pass. I limit my ambition, in these light pages, to speaking of those things of which I have personally received an impression; and I have no such impression of the defense of La Rochelle. The hotel de ville is a pretty little building, in the style of the Renaissance of Francis I.; but it has left much of its interest in the hands of the restorers. It has been "done up" without mercy; its natural place would be at Rochelle the New. A sort of battlemented curtain, flanked with turrets, divides it from the street and contains a low door (a low door in a high wall is always felicitous), which admits you to an inner court, where you discover the face of the building. It has statues set into it, and is raised upon a very low and very deep arcade. The principal function of the deferential old portress who conducts you over the place is to call your attention to the indented table of Jean Guiton; but she shows you other objects of interest besides. The interior is absolutely new and extremely sumptuous, abounding in tapestries, upholstery, morocco, velvet, satin. This is especially the case with a really beautiful *grande salle*, where, surrdunded with the most expensive upholstery, the mayor holds his official receptions. (So at least, said my worthy portress.) The mayors of La Rochelle appear to have changed a good deal since the days of the grim Guiton; but these evidences of municipal splendor are interesting for the light they throw on French manners. Imagine the mayor of an English or an American town of twenty thousand inhabitants holding magisterial soirees in the town-hall! The said *grande salle*, which is unchanged in form and its larger features, is, I believe, the room in which the Rochelais debated as to whether they should shut themselves up, and decided in the affirmative. The table and chair of Jean Guiton have been restored, like everything else, and are very elegant and coquettish pieces of furniture, incongruous relics of a season of starvation and blood. I believe that Protestantism is somewhat shrunken to-day at La Rochelle, and has taken refuge mainly in. the *haute societe* and in a single place of worship. There was nothing particular to remind me of its supposed austerity as, after leaving the hotel de ville, I walked along the empty portions and cut out of the Tour de l'Horloge, which I have already mentioned. If I stopped and looked up at this venerable monument, it was not to ascertain the hour, for I foresaw that I should have more time at La Rochelle than I knew what to do with; but because its high, gray, weather-beaten face was an obvious subject for a sketch. The little port, which has two basins, and is accessible only to vessels of light tonnage, had a certain gayety and as much local color as you please. Fisher folk of picturesque type were strolling about, most of them Bretons; several of the men with handsome, simple faces, not at all brutal, and with a splendid brownness, the golden-brown color, on cheek and beard, that you see on an old Venetian sail. It was a squally, showery day, with sudden drizzles of sunshine; rows of rich-toned fishing-smacks were drawn up along the quays. The harbor is effective to the eye by reason of three battered old towers which, at different points, overhang it and look infinitely weatherwashed and sea-silvered. The most striking of these, the Tour de la Lanterne, is a big gray mass, of the fifteenth century, flanked with turrets and crowned with a Gothic steeple. I found it was called by the people of the place the Tour des Quatre Sergents, though I know not what connection it has with the touching history of the four young sergeants of the garrison of La Rochelle, who were arrested in 1821 as conspirators against the Government of the Bourbons, and executed, amid general indignation, in Paris in the following year. The quaint little walk, with its label of Rue sur les Murs, to which one ascends from beside the Grosse Horloge, leads to this curious Tour de la Lanterne and passes under it. This walk has the top of the old town-wall, toward the sea, for a parapet on one side, and is bordered on the other with decent but irregular little tenements of fishermen, where brown old women, whose caps are as white as if they were painted, seem chiefly in possession. In this direction there is a very pretty stretch of shore, out of the town, through the fortifications (which are Vauban's, by the way); through, also, a diminutive public garden or straggling shrubbery, which edges the water and carries its stunted verdure as far as a big Etablissernent des Bains. It was too late in the year to bathe, and the Etablissement had the bankrupt aspect which belongs to such places out of the season; so I turned my back upon it, and gained, by a circuit in the course of which there were sundry water-side items to observe, the other side of the cheery little port, where there is a long breakwater and a still longer sea-wall, on which I walked awhile, to inhale the strong, salt breath of the Bay of Biscay. La Rochelle serves, in the months of July and August, as a *station de bains* for a modest provincial society; and, putting aside the question of inns, it must be charming on summer afternoons.

17

It is an injustice to Poitiers to approach her by night, as I did some three hours after leaving La Rochelle; for what Poitiers has of best, as they would say at Poitiers, is the appearance she presents to the arriving stranger who puts his head out of the window of the train. I gazed into the gloom from such an aperture before we got into the station, for I remembered the impression received on another occasion; but I saw nothing save the universal night, spotted here and there with an ugly railway lamp. It was only as I departed, the following day, that I assured myself that Poitiers still makes something of the figure she ought on the summit of her considerable bill. I have a kindness for any little group of towers, any cluster of roofs and chimneys, that lift themselves from an eminence over which a long road ascends in zigzags; such a picture creates for the moment a presumption that you are in Italy, and even leads you to believe that if you mount the winding road you will come to an old town-wall, an expanse of creviced brownness, and pass under a gateway surmounted by the arms of a mediaeval despot. Why I should find it a pleasure, in France, to imagine myself in Italy, is more than I can say; the illusion has never lasted long enough to be analyzed. From the bottom of its perch Poitiers looks large and high; and indeed, the evening I reached it, the interminable climb of the omnibus of the hotel I had selected, which I found at the station, gave me the measure of its commanding position. This hotel, "magnifique construction ornee de statues," as the Guide-Joanne, usually so reticent, takes the trouble to announce, has an omnibus, and, I suppose, has statues, though I didn't perceive them; but it has very little else save immemorial accumulations of dirt. It is magnificent, if you will, but it is not even relatively proper; and a dirty inn has always seemed to me the dirtiest of human things, it has so many opportunities to betray itself.

Poiters covers a large space, and is as crooked and straggling as you please; but these advantages are not accompanied with any very salient features or any great wealth of architecture. Although there are few picturesque houses, however, there are two or three curious old churches. Notre Dame la Grande, in the market-place, a small Romanesque structure of the twelfth century, has a most interesting and venerable exterior. Composed, like all the churches of Poitiers, of a light brown stone with a yellowish tinge, it is covered with primitive but ingenious sculptures, and is really an impressive monument. Within, it has lately been daubed over with the most hideous decorative painting that was ever inflicted upon passive pillars and indifferent vaults. This battered yet coherent little edifice has the touching look that resides in everything supremely old: it has arrived at the age at which such things cease to feel the years; the waves of time have worn its edges to a kind of patient dullness; there is something mild and smooth, like the stillness, the deafness, of an octogenarian, even in its rudeness of ornament, and it has become insensible to differences of a century or two. The cathedral interested me much less than Our Lady the Great, and I have not the spirit to go into statistics about it. It is not statistical to say that the cathedral stands half-way down the hill of Poitiers, in a quiet and grass-grown *place*, with an approach of crooked lanes and blank garden-walls, and that its most striking dimension is the width of its facade. This width is extraordinary, but it fails, somehow, to give nobleness to the edifice, which looks within (Murray makes the remark) like a large public hall. There are a nave and two aisles, the latter about as high as the nave; and there are some very fearful modern pictures, which you may see much better than you usually see those specimens of the old masters that lurk in glowing side-chapels, there being no fine old glass to diffuse a kindly gloom. The sacristan of the cathedral showed me something much better than all this bright bareness; he led me a short distance out of it to the small Temple de Saint-Jean, which is the most curious object at Poitiers. It is an early Christian chapel, one of the earliest in France; originally, it would seem, that is, in the sixth or seventh century, a baptistery, but converted into a church while the Christian era was still comparatively young. The Temple de Saint-Jean is therefore a monument even more venerable than Notre

Dame la Grande, and that numbness of age which I imputed to Notre Dame ought to reside in still larger measure in its crude and colorless little walls. I call them crude, in spite of their having been baked through by the centuries, only because, although certain rude arches and carvings are let into them, and they are surmounted at either end with a small gable, they have (so far as I can remember) little fascination of surface. Notre Dame is still expressive, still pretends to be alive; but the Temple has delivered its message, and is completely at rest. It retains a kind of atrium, on the level of the street, from which you descend to the original floor, now uncovered, but buried for years under a false bottom. A semicircular apse was, apparently at the time of its conversion into a church, thrown out from the east wall. In the middle is the cavity of the old baptismal font. The walls and vaults are covered with traces of extremely archaic frescos, attributed, I believe, to the twelfth century. These vague, gaunt, staring fragments of figures are, to a certain extent, a reminder of some of the early Christian churches in Rome; they even faintly recalled to me the great mosaics of Ravenna. The Temple de Saint-Jean has neither the antiquity nor the completeness of those extraordinary monuments, nearly the most impressive in Europe; but, as one may say, it is very well for Poitiers.

Not far from it, in a lonely corner which was animated for the moment by the vociferations of several old, women who were selling tapers, presumably for the occasion of a particular devotion, is the graceful Romanesque church erected in the twelfth century to Saint Radegonde, a lady who found means to be a saint even in the capacity of a Merovingian queen. It bears a general resemblance to Notre Dame la Grande, and, as I remember it, is corrugated in somewhat the same manner with porous-looking carvings; but I confess that what I chiefly recollect is the row of old women sitting in front of it, each with a tray of waxen tapers in her lap, and upbraiding me for my neglect of the opportunity to offer such a tribute to the saint. I know not whether this privilege is occasional or constant; within the church there was no appearance of a festival, and I see that the nameday of Saint Radegonde occurs in August, so that the importunate old women sit there always, perhaps, and deprive of its propriety the epithet I just applied to this provincial corner. In spite of the old women, however, I suspect that the place is lonely; and indeed it is perhaps the old women that have made the desolation.

The lion of Poitiers, in the eyes of the natives, is doubtless the Palais de Justice, in the shadow of which the statue-guarded hotel, just mentioned, erects itself; and the gem of the court-house, which has a prosy modern front, with pillars and a high flight of steps, is the curious *salle des pas perdus*, or central hall, out of which the different tribunals open. This is a feature of every French court-house, and seems the result of a conviction that a palace of justice the French deal in much finer names than we should be in some degree palatial. The great hall at Poitiers has a long pedigree, as its walls date back to the twelfth century, and its open wooden roof, as well as the remarkable trio of chimney-pieces at the right end of the room as you enter, to the fifteenth. The three tall fireplaces, side by side, with a delicate gallery running along the top of them, constitute the originality of this ancient chamber, and make one think of the groups that must formerly have gathered there, of all the wet boot-soles, the trickling doublets, the stiffened fingers, the rheumatic shanks, that must have been presented to such an incomparable focus of heat. To-day, I am afraid, these mighty hearts are forever cold; justice it probably administered with the aid of a modern *calorifere*, and the walls of the palace are perforated with regurgitating tubes. Behind and above the gallery that surmounts the three fireplaces are high Gothic windows, the tracery of which masks, in some sort, the chimneys; and in each angle of this and of the room to the right and left of the trio of chimneys, is all open-work spiral staircase, ascending to I forget where; perhaps to the roof of the edifice. This whole side of the *salle* is very lordly, and seems to express an unstinted hospitality, to extend the friendliest of all invitations, to bid the whole world come and get warm. It was the invention of John, Duke of Berry and Count of Poitou, about 1395. I give this information on the authority of the GuideJoanne, from which source I gather much other curious learning; for instance, that it was in this building, when it had surely a very different front, that Charles VII. was proclaimed king, in 1422; and that here Jeanne Darc was subjected, in 1429, to the inquisition of certain doctors and matrons.

The most charming thing at Poitiers is simply the Promenade de Blossac, a small public garden at one end of the flat top of the hill. It has a happy look of the last century (having been arranged at that period), and a beautiful sweep of view over the surrounding country, and especially of the course of the little river Clain, which winds about a part of the base of the big mound of Poitiers. The limit of this dear little garden is formed, on the side that turns away from the town, by the rampart erected in the fourteenth century, and by its big semicircular bastions. This rampart, of great length, has a low parapet; you look over it at the charming little vegetable-gardens with which the base of the hill appears exclusively to be garnished. The whole prospect is delightful, especially the details of the part just under the walls, at the end of the walk. Here the river makes a shining twist, which a painter might have invented, and the side of the hill is terraced into several ledges, a sort of tangle of small blooming patches and little pavilions with peaked roofs and green shutters. It is idle to attempt to reproduce all this in words; it should be reproduced only in water-colors. The reader, however, will already have remarked that disparity in these ineffectual pages, which are pervaded by the attempt to sketch without a palette or brushes. He will doubtless, also, be struck with the groveling vision which, on such a spot as the ramparts of Poitiers, peoples itself with carrots and cabbages rather than with images of the Black Prince and the captive king. I am not sure that in looking out from the Promenade de Blossac you command the old battle-field; it is enough that it was not far off, and that the great rout of Frenchmen poured into the walls of Poitiers, leaving on the ground a number of the fallen equal to the little army (eight thousand) of the invader. I did think of the battle. I wondered, rather helplessly, where it had taken place; and I came away (as the reader will see from the preceding sentence) without finding out. This indifference, however, was a result rather of a general dread of military topography than of a want of admiration of this particular victory, which I have always supposed to be one of the most brilliant on record. Indeed, I should be almost ashamed, and very much at a loss, to say what light it was that this glorious day seemed to me to have left forever on the horizon, and why the very name of the place had always caused my blood gently to tingle. It is carrying the feeling of race to quite inscrutable lengths when a vague American permits himself an emotion because more than five centuries ago, on French soil, one rapacious Frenchman got the better of another. Edward was a Frenchman as well as John, and French were the cries that urged each of the hosts to the fight. French is the beautiful motto graven round the image of the Black Prince, as he lies forever at rest in the choir of Canterbury: *a la mort ne pensai-je mye.* Nevertheless, the victory of Poitiers declines to lose itself in these considerations; the sense of it is a part of our heritage, the joy of it a part of our imagination, and it filters down through centuries and migrations till it titillates a New Yorker who forgets in his elation that he happens at that moment to be enjoying the hospitality of France. It was something done, I know not how justly, for England; and what was done in the fourteenth century for England was done also for New York.

18

If it was really for the sake of the Black Prince that I had stopped at Poitiers (for my prevision of Notre Dame la Grande and of the little temple of St. John was of the dimmest), I ought to have stopped at Angouleme for the sake of David and Eve Sechard, of Lucien de Rubempre and of Madame de Bargeton, who when she wore a *toilette etudiee* sported a Jewish turban ornamented with an Eastern brooch, a scarf of gauze, a necklace of cameos, and a robe of "painted muslin," whatever that may be; treating herself to these luxuries out of an income of twelve thousand francs. The persons I have mentioned have not that vagueness of identity which is the misfortune of historical characters; they are real, supremely real, thanks to their affiliation to the great Balzac, who had invented an artificial reality which was as much better than the vulgar article as mock-turtle soup is than the liquid it emulates. The first time I read "Les Illusions Perdues" I should have refused to believe that I was capable of passing the old capital of Anjou without alighting to visit the Houmeau. But we never know what we are capable of till we are tested, as I reflected when I found myself looking back at Angouleme from the window of the train, just after we had emerged from the long tunnel that passes under the town. This tunnel perforates the hill on which, like Poitiers, Angouleme rears itself, and which gives it an elevation still greater than that of Poitiers. You may have a tolerable look at the cathedral without leaving the railway-carriage; for it stands just above the tunnel, and is exposed, much foreshortened, to the spectator below. There is evidently a charming walk round the plateau of the town, commanding those pretty views of which Balzac gives an account. But the train whirled me away, and these are my only impressions. The truth is that I had no need, just at that moment, of putting myself into communication with Balzac; for opposite to me in the compartment were a couple of figures almost as vivid as the actors in the "Comedie Humaine." One of these was a very genial and dirty old priest, and the other was a reserved and concentrated young monk, the latter (by which I mean a monk of any kind) being a rare sight to-day in France. This young man, indeed, was mitigatedly monastic. He had a big brown frock and cowl, but he had also a shirt and a pair of shoes; he had, instead of a hempen scourge round his waist, a stout leather thong, and he carried with him a very profane little valise. He also read, from beginning to end, the "Figaro" which the old priest, who had done the same, presented to him; and he looked altogether as if, had he not been a monk, he would have made a distinguished officer of engineers. When he was not reading the "Figaro" he was conning his breviary or answering, with rapid precision and with a deferential but discouraging dryness, the frequent questions of his companion, who was of quite another type. This worthy had a bored, good-natured, unbuttoned, expansive look; was talkative, restless, almost disreputably human. He was surrounded by a great deal of small luggage, and had scattered over the carriage his books, his papers, the fragments of his lunch, and the contents of an extraordinary bag, which he kept beside him a kind of secular reliquary and which appeared to contain the odds and ends of a lifetime, as he took from it successively a pair of slippers, an old padlock (which evidently didn't belong to it), an opera-glass, a collection of almanacs, and a large sea-shell, which he very carefully examined. I think that if he had not been afraid of the young monk, who was so much more serious than he, he would have held the shell to his ear, like a child. Indeed, he was a very childish and delightful old priest, and his companion evidently thought him most frivolous. But I liked him the better of the two. He was not a country cure, but an ecclesiastic of some rank, who had seen a good deal both of the church and of the world; and if I too had not been afraid of his colleague, who read the "Figaro" as seriously as if it had been an encyclical, I should have entered into conversation with him.

All this while I was getting on to Bordeaux, where I permitted myself to spend three days. I am afraid I have next to nothing to show for them, and that there would be little profit in lingering on this episode, which is the less to be justified as I had in former years examined Bordeaux attentively enough. It contains a very good hotel, an hotel not good enough, however, to keep you there for its own sake. For the rest, Bordeaux is a big, rich, handsome, imposing commercial town, with long rows of fine old eighteenth century houses, which overlook the yellow Garonne. I have spoken of the quays of Nantes as fine, but those of Bordeaux have a wider sweep and a still more architectural air. The appearance of such a port as this makes the Anglo-Saxon tourist blush for the sordid water-fronts of Liverpool and New York, which, with their larger activity, have so much more reason to be stately. Bordeaux gives a great impression of prosperous industries, and suggests delightful ideas, images of prune-boxes and bottled claret. As the focus of distribution of the best wine in the world, it is indeed a sacred city, dedicated to the worship of Bacchus in the most discreet form. The country all about it is covered with precious vineyards, sources of fortune to their owners and of satisfaction to distant consumers; and as you look over to the hills beyond the Garonne you see them in the autumn sunshine, fretted with the rusty richness of this or that immortal *clos*. But the principal picture, within the town, is that of the vast curving quays, bordered with houses that look like the *hotels* of farmers-general of the last century, and of the wide, tawny river, crowded with shipping and spanned by the largest of bridges. Some of the types on the water-side are of the sort that arrest a sketcher, figures of stalwart, brown-faced Basques, such as I had seen of old in great numbers at Biarritz, with their loose circular caps, their white sandals, their air of walking for a wager. Never was a tougher, a harder race. They are not mariners, nor watermen, but, putting questions of temper aside, they are the best possible dock-porters. "Il s'y fait un commerce terrible," a *douanier* said to me, as he looked up and down the interminable docks; and such a place has indeed much to say of the wealth, the capacity for production, of France, the bright, cheerful, smokeless industry of the wonderful country which produces, above all, the agreeable things of life, and turns even its defeats and revolutions into gold. The whole town has an air of almost depressing opulence, an appearance which culminates in the great *place* which surrounds the Grand-Theatre, an establishment in the highest style, encircled with columns, arcades, lamps, gilded cafes. One feels it to be a monument to the virtue of the well-selected bottle. If I had not forbidden myself to linger, I should venture to insist on this, and, at the risk of being considered fantastic, trace an analogy between good claret and the best qualities of the French mind; pretend that there is a taste of sound Bordeaux in all the happiest manifestations of that fine organ, and that, correspondingly, there is a touch of French reason, French completeness, in a glass of Pontet-Canet. The danger of such an excursion would lie mainly in its being so open to the reader to take the ground from under my feet by saying that good claret doesn't exist. To this I should have no reply whatever. I should be unable to tell him where to find it. I certainly didn't find it at Bordeaux, where I drank a most vulgar fluid; and it is of course notorious that a large part of mankind is occupied in vainly looking for it. There was a great pretence of putting it forward at the Exhibition which was going on at Bordeaux at the time of my visit, an "exposition philomathique," lodged in a collection of big temporary buildings in the Allees d'Or1eans, and regarded by the Bordelais for the moment as the most brilliant feature of their city. Here were pyramids of bottles, mountains of bottles, to say nothing of cases and cabinets of bottles. The contemplation of these glittering tiers was of course not very convincing; and indeed the whole arrangement struck me as a high impertinence. Good wine is not an optical pleasure, it is an inward emotion; and if there was a chamber of degustation on the premises, I failed to discover it. It was not in the search for it, indeed, that I spent half an hour in this bewildering bazaar. Like all "expositions," it seemed to me to be full of ugly things, and gave one a portentous idea of the quantity of rubbish that man carries with him on his course through the ages. Such an amount of luggage for a journey after all so short! There were no individual objects; there was nothing but dozens and hundreds, all machine-made and expressionless, in spite of the repeated grimace, the conscious smartness, of "the last new thing," that was stamped on all of them. The fatal facility, of the French *article* becomes at last as irritating as the refrain of a popular song. The poor "Indiens Galibis" struck me as really more interesting, a group of stunted savages who formed one of the attractions of the place, and were confined in a pen in the open air, with a rabble of people pushing and squeezing, hanging over the barrier, to look at them. They had no grimace, no pretension to be new, no desire to catch your eye. They looked at their visitors no more than they looked at each other, and seemed ancient, indifferent, terribly bored.

19

There is much entertainment in the journey through the wide, smiling garden of Gascony; I speak of it as I took it in going from Bordeaux to Toulouse. It is the south, quite the south, and had for the present narrator its full measure of the charm he is always determined to find in countries that may even by courtesy be said to appertain to the sun. It was, moreover, the happy and genial view of these mild latitudes, which, Heaven knows, often have a dreariness of their own; a land teeming with corn and wine, and speaking everywhere (that is, everywhere the phylloxera had not laid it waste) of wealth and plenty. The road runs constantly near the Garonne, touching now and then its slow, brown, rather sullen stream, a sullenness that encloses great dangers and disasters. The traces of the horrible floods of 1875 have disappeared, and the land smiles placidly enough while it waits for another immersion. Toulouse, at the period I speak of, was up to its middle (and in places above it) in water, and looks still as if it had been thoroughly soaked, as if it had faded and shriveled with a long steeping. The fields and copses, of course, are more forgiving. The railway line follows as well the charming Canal du Midi, which is as pretty as a river, barring the straightness, and here and there occupies the foreground, beneath a screen of dense, tall trees, while the Garonne takes a larger and more irregular course a little way beyond it. People who are fond of canals and, speaking from the pictorial standpoint, I hold the taste to be most legitimate will delight in this admirable specimen of the class, which has a very interesting history, not to be narrated here. On the other side of the road (the left), all the way, runs a long, low line of hills, or rather one continuous hill, or perpetual cliff, with a straight top, in the shape of a ledge of rock, which might pass for a ruined wall. I am afraid the reader will lose patience with my habit of constantly referring to the landscape of Italy, as if that were the measure of the beauty of every other. Yet I am still more afraid that I cannot apologize for it, and must leave it in its culpable nakedness. It is an idle habit; but the reader will long since have discovered that this was an idle journey, and that I give my impressions as they came to me. It came to me, then, that in all this view there was something transalpine with a greater smartness and freshness and much less elegance and languor. This impression was occasionally deepened by the appearance, on the long eminence of which I speak, of a village, a church, or a chateau, which seemed to look down at the plain from over the ruined wall. The perpetual vines, the bright-faced flat-roofed houses, covered with tiles, the softness and sweetness of the light and air, recalled the prosier portions of the Lombard plain. Toulouse itself has a little of this Italian expression, but not enough to give a color to its dark, dirty, crooked streets, which are irregular without being eccentric, and which, if it were not for the, superb church of Saint-Sernin, would be quite destitute of monuments.

I have already alluded to the way in which the names of certain places impose themselves on the mind, and I must add that of Toulouse to the list of expressive appellations. It certainly evokes a vision, suggests something highly *meridional*. But the city, it must be confessed, is less pictorial than the word, in spite of the Place du Capitole, in spite of the quay of the Garonne, in spite of the curious cloister of the old museum. What justifies the images that are latent in the word is not the aspect, but the history, of the town. The hotel to which the well-advised traveler will repair stands in a corner of the Place du Capitole, which is the heart and centre of Toulouse, and which bears a vague and inexpensive resemblance to Piazza Castello at Turin. The Capitol, with a wide modern face, occupies one side, and, like the palace at Turin, looks across at a high arcade, under which the hotels, the principal shops, and the lounging citizens are gathered. The shops are probably better than the Turinese, but the people are not so good. Stunted, shabby, rather vitiated looking, they have none of the personal richness of the sturdy Piedmontese; and I will take this occasion to remark that in the course of a journey of several weeks in the French provinces I rarely encountered a well-dressed male. Can it be possible the republics are unfavorable to a certain attention to one's boots and one's beard? I risk this somewhat futile inquiry because the proportion of neat coats and trousers seemed to be about the same in France and in my native land. It was notably lower than in England and in Italy, and even warranted the supposition that most good provincials have their chin shaven and their boots blacked but once a week. I hasten to add, lest my observation should appear to be of a sadly superficial character, that the manners and conversation of these gentlemen bore (whenever I had occasion to appreciate them) no relation to the state of their chin and their boots. They were almost always marked by an extreme amenity. At Toulouse there was the strongest temptation to speak to people, simply for the entertainment of hearing them reply with that curious, that fascinating accent of the Languedoc, which appears to abound in final consonants, and leads the Toulousains to say *bien-g* and *maison-g*, like Englishmen learning French. It is as if they talked with their teeth rather than with their tongue. I find in my note-book a phrase in regard to Toulouse which is perhaps a little ill-natured, but which I will transcribe as it stands: "The oddity is that the place should be both animated and dull. A big, brown-skinned population, clattering about in a flat, tortuous town, which produces nothing whatever that I can discover. Except the church of Saint Sernin and the fine old court of the Hotel d'Assezat, Toulouse has no architecture; the houses are for the most part of brick, of a grayish-red color, and have no particular style. The brick-work of the place is in fact very poor, inferior to that of the north Italian towns, and quite wanting in the richness of tone which this homely material takes on

83

in the damp climates of the north." And then my note-book goes on to narrate a little visit to the Capitol, which was soon made, as the building was in course of repair and half the rooms were closed.

20

The history of Toulouse is detestable, saturated with blood and perfidy; and the ancient custom of the Floral Games, grafted upon all sorts of internecine traditions, seems, with its false pastoralism, its mock chivalry, its display of fine feelings, to set off rather than to mitigate these horrors. The society was founded in the fourteenth century, and it has held annual meetings ever since, meetings at which poems in the fine old *langue d'oc* are declaimed and a blushing laureate is chosen. This business takes place in the Capitol, before the chief magistrate of the town, who is known as the *capitoul*, and of all the pretty women as well, a class very numerous at Toulouse. It was impossible to have a finer person than that of the portress who pretended to show me the apartments in which the Floral Games are held; a big, brown, expansive woman, still in the prime of life, with a speaking eye, an extraordinary assurance, and a pair of magenta stockings, which were inserted into the neatest and most polished little black sabots, and which, as she clattered up the stairs before me, lavishly displaying them, made her look like the heroine of an *opera-bouffe.* Her talk was all in *n*'s, *g*'s, and *d*'s, and in mute *e*'s strongly accented, as *autre, theatre, splendide,* the last being an epithet she applied to everything the Capitol contained, and especially to a horrible picture representing the famous Clemence Isaure, the reputed foundress of the poetical contest, presiding on one of these occasions. I wondered whether Clemence Isaure had been anything like this terrible Toulousaine of to-day, who would have been a capital figure-head for a floral game. The lady in whose honor the picture I have just mentioned was painted is a somewhat mythical personage, and she is not to be found in the "Biographie Universelle." She is, however, a very graceful myth; and if she never existed, her statue does, at least, a shapeless effigy, transferred to the Capitol from the so-called tomb of Clemence in the old church of La Daurade. The great hall in which the Floral Games are held was encumbered with scaffoldings, and I was unable to admire the long series of busts of the bards who have won prizes and the portraits of all the capitouls of Toulouse. As a compensation I was introduced to a big bookcase, filled with the poems that have been crowned since the days of the troubadours (a portentous collection), and the big butcher's knife with which, according to the legend, Henry, Duke of Montmorency, who had conspired against the great cardinal with Gaston of Orleans and Mary de Medici, was, in 1632, beheaded on this spot by the order of Richelieu. With these objects the interest of the Capitol was exhausted. The building, indeed, has not the grandeur of its name, which is a sort of promise that the visitor will find some sensible embodiment of the old Roman tradition that once flourished in this part of France. It is inferior in impressiveness to the other three famous Capitols of the modern world, that of Rome (if I may call the present structure modern) and those of Washington and Albany!

The only Roman remains at Toulouse are to be found in the museum, a very interesting establishment, which I was condemned to see as imperfectly as I had seen the Capitol. It was being rearranged; and the gallery of paintings, which is the least interesting feature, was the only part that was not upside-down. The pictures are mainly of the modern French school, and I remember nothing but a powerful, though disagreeable specimen of Henner, who paints the human body, and paints it so well, with a brush dipped in blackness; and, placed among the paintings, a bronze replica of the charming young David of Mercie. These things have been set out in the church of an old monastery, long since suppressed, and the rest of the collection occupies the cloisters. These are two in number, a small one, which you enter first from the street, and a very vast and elegant one beyond it, which with its light Gothic arches and slim columns (of the fourteenth century), its broad walk its little garden, with old tombs and statues in the centre, is by far the most picturesque, the most sketchable, spot in Toulouse. It must be doubly so when the Roman busts, inscriptions, slabs and sarcophagi, are ranged along the walls; it must indeed (to compare small things with great, and as the judicious Murray remarks) bear a certain resemblance to the Campo Santo at Pisa. But these things are absent now; the cloister is a litter of confusion, and its treasures have been stowed away, confusedly, in sundry inaccessible rooms. The custodian attempted to console me by telling me that when they are exhibited again it will be on a scientific basis, and with an order and regularity of which they were formerly innocent. But I was not consoled. I wanted simply the spectacle, the picture, and I didn't care in the least for the

85

classification. Old Roman fragments, exposed to light in the open air, under a southern sky, in a quadrangle round a garden, have an immortal charm simply in their general effect; and the charm is all the greater when the soil of the very place has yielded them up.

21

My real consolation was an hour I spent in Saint Sernin, one of the noblest churches in southern France, and easily the first among those of Toulouse. This great structure, a masterpiece of twelfth-century Romanesque, and dedicated to Saint Saturninus, the Toulousains have abbreviated, is, I think, alone worth a journey to Toulouse. What makes it so is the extraordinary seriousness of its interior; no other term occurs to me as expressing so well the character of its clear gray nave. As a general thing, I do not favor the fashion of attributing moral qualities to buildings; I shrink from talking about tender porticos and sincere campanili; but I find I cannot get on at all without imputing some sort of morality to Saint Sernin. As it stands to-day, the church has been completely restored by Viollet-le-Duc. The exterior is of brick, and has little charm save that of a tower of four rows of arches, narrowing together as they ascend. The nave is of great length and height, the barrel-roof of stone, the effect of the round arches and pillars in the triforium especially fine. There are two low aisles on either side. The choir is very deep and narrow; it seems to close together, and looks as if it were meant for intensely earnest rites. The transepts are most noble, especially the arches of the second tier. The whole church is narrow for its length, and is singularly complete and homogeneous. As I say all this, I feel that I quite fail to give an impression of its manly gravity, its strong proportions or of the lonesome look of its renovated stones as I sat there while the October twilight gathered. It is a real work of art, a high conception. The crypt, into which I was eventually led captive by an importunate sacristan, is quite another affair, though indeed I suppose it may also be spoken of as a work of art. It is a rich museum of relics, and contains the head of Saint Thomas Aquinas, wrapped up in a napkin and exhibited in a glass case. The sacristan took a lamp and guided me about, presenting me to one saintly remnant after another. The impression was grotesque, but sorne of the objects were contained in curious old cases of beaten silver and brass; these things, at least, which looked as if they had been transmitted from the early church, were venerable. There was, however, a kind of wholesale sanctity about the place which overshot the mark; it pretends to be one of the holiest spots in the world. The effect is spoiled by the way the sacristans hang about and offer to take you into it for ten sous, I was accosted by two and escaped from another, and by the familiar manner in which you pop in and out. This episode rather broke the charm of Saint-Sernin, so that I took my departure and went in search of the cathedral. It was scarcely worth finding, and struck me as an odd, dislocated fragment. The front consists only of a portal, beside which a tall brick tower, of a later period, has been erected. The nave was wrapped in dimness, with a few scattered lamps. I could only distinguish an immense vault, like a high cavern, without aisles. Here and there in the gloom was a kneeling figure; the whole place was mysterious and lop-sided. The choir was curtained off; it appeared not to correspond with the nave, that is, not to have the same axis. The only other ecclesiastical impression I gathered at Toulouse came to me in the church of La Daurade, of which the front, on the quay by the Garonne, was closed with scaffoldings; so that one entered it from behind, where it is completely masked by houses, through a door which has at first no traceable connection with it. It is a vast, high, modernized, heavily decorated church, dimly lighted at all times, I should suppose, and enriched by the shades of evening at the time I looked into it. I perceived that it consisted mainly of a large square, beneath a dome, in the centre of which a single person a lady was praying with the utmost absorption. The manner of access to the church interposed such an obstacle to the outer profanities that I had a sense of intruding, and presently withdrew, carrying with me a picture of the, vast, still interior, the gilded roof gleaming in the twilight, and the solitary worshipper. What was she praying for, and was she not almost afraid to remain there alone?

For the rest, the picturesque at Toulouse consists principally of the walk beside the Garonne, which is spanned, to the faubourg of Saint-Cyprien, by a stout brick bridge. This hapless suburb, the baseness of whose site is noticeable, lay for days under the water at the time of the last inundations. The Garonne had almost mounted to the roofs of the houses, and the place continues to present a blighted, frightened look. Two or three persons, with whom I had some conversation, spoke of that time as a memory of horror. I have not done with my Italian comparisons; I shall never have done with them. I am therefore free to say that in the way in which Toulouse looks out on the Garonne there was something that reminded me vaguely of the way in which Pisa looks out on the Arno. The red-faced houses all of brick along the quay have a mixture of brightness and shabbiness, as well as the fashion of the open *loggia* in the top story. The river, with another bridge or two, might be the Arno, and the buildings on the other side of it a hospital, a suppressed convent dip their feet into it with real southern cynicism. I have spoken of the old Hotel d'Assezat as the best house at Toulouse; with the exception of the cloister of the museum, it is the only "bit" I remember. It has fallen from the state of a noble residence of the sixteenth century to that of a warehouse and a set of offices; but a certain dignity lingers in its melancholy court, which is divided from the street by a gateway that is still imposing, and in which a clambering vine and a red Virginia creeper were suspended to the rusty walls of brick stone.

The most interesting house at Toulouse is far from being the most striking. At the door of No. 50 Rue des Filatiers, a featureless, solid structure, was found hanging, one autumn evening, the body of the young Marc-Antoine Calas, whose ill-inspired suicide was to be the first act of a tragedy so horrible. The fanaticism aroused in the townsfolk by this incident; the execution by torture of Jean Calas, accused as a Protestant of having hanged his son, who had gone over to the Church of Rome; the ruin of the family; the claustration of the daughters; the flight of the widow to Switzerland; her introduction to Voltaire; the excited zeal of that incomparable partisan, and the passionate persistence with which, from year to year, he pursued a reversal of judgment, till at last he obtained it, and devoted the tribunal of Toulouse to execration and the name of the victims to lasting wonder and pity, these things form part of one of the most interesting and touching episodes of the social history of the eighteenth century. The story has the fatal progression, the dark rigidity, of one of the tragic dramas of the Greeks. Jean Calas, advanced in life, blameless, bewildered, protesting. his innocence, had been broken on the wheel; and the sight of his decent dwelling, which brought home to me all that had been suffered there, spoiled for me, for half an hour, the impression of Toulouse.

22

I spent but a few hours at Carcassonne; but those hours had a rounded felicity, and I cannot do better than transcribe from my note-book the little record made at the moment. Vitiated as it may be by crudity and incoherency, it has at any rate the freshness of a great emotion. This is the best quality that a reader may hope to extract from a narrative in which "useful information" and technical lore even of the most general sort are completely absent. For Carcassonne is moving, beyond a doubt; and the traveler who, in the course of a little tour in France, may have felt himself urged, in melancholy moments, to say that on the whole the disappointments are as numerous as the satisfactions, must admit that there can be nothing better than this.

The country, after you leave Toulouse, continues to be charming; the more so that it merges its flatness in the distant Cevennes on one side, and on the other, far away on your right, in the richer range of the Pyrenees. Olives and cypresses, pergolas and vines, terraces on the roofs of houses, soft, iridescent mountains, a warm yellow light, what more could the difficult tourist want? He left his luggage at the station, warily determined to look at the inn before committing himself to it. It was so evident (even to a cursory glance) that it might easily have been much better that he simply took his way to the town, with the whole of a superb afternoon before him. When I say the town, I mean the towns; there being two at Carcassonne, perfectly distinct, and each with excellent claims to the title. They have settled the matter between them, however, and the elder, the shrine of pilgrimage, to which the other is but a stepping-stone, or even, as I may say, a humble door-mat, takes the name of the Cite. You see nothing of the Cite from the station; it is masked by the agglomeration of the *ville-basse*, which is relatively (but only relatively) new. A wonderful avenue of acacias leads to it from the station, leads past, rather, and conducts you to a little high-backed bridge over the Aude, beyond which, detached and erect, a distinct mediaeval silhouette, the Cite presents itself. Like a rival shop, on the invidious side of a street, it has "no connection" with the establishment across the way, although the two places are united (if old Carcassonne may be said to be united to anything) by a vague little rustic faubourg. Perched on its solid pedestal, the perfect detachment of the Cite is what first strikes you. To take leave, without delay, of the *ville-basse*, I may say that the splendid acacias I have mentioned flung a summerish dusk over the place, in which a few scattered remains of stout walls and big bastions looked venerable and picturesque. A little boulevard winds round the town, planted with trees and garnished with more benches than I ever saw provided by a soft-hearted municipality. This precinct had a warm, lazy, dusty, southern look, as if the people sat out-of-doors a great deal, and wandered about in the stillness of summer nights. The figure of the elder town, at these hours, must be ghostly enough on its neighboring hill. Even by day it has the air of a vignette of Gustave Dore, a couplet of Victor Hugo. It is almost too perfect, as if it were an enormous model, placed on a big green table at a museum. A steep, paved way, grass-grown like all roads where vehicles never pass, stretches up to it in the sun. It has a double enceinte, complete outer walls and complete inner (these, elaborately fortified, are the more curious); and this congregation of ramparts, towers, bastions, battlements, barbicans, is as fantastic and romantic as you please. The approach I mention here leads to the gate that looks toward Toulouse, the Porte de l'Aude. There is a second, on the other side, called, I believe, the Porte Narbonnaise, a magnificent gate, flanked with towers thick and tall, defended by elaborate outworks; and these two apertures alone admit you to the place, putting aside a small sally-port, protected by a great bastion, on the quarter that looks toward the Pyrenees.

As a votary, always, in the first instance, of a general impression, I walked all round the outer enceinte, a process on the very face of it entertaining. I took to the right of the Porte de l'Aude, without entering it, where the old moat has been filled in. The filling-in of the moat has created a grassy level at the foot of the big gray towers, which, rising at frequent intervals, stretch their stiff curtain of stone from point to point. The curtain drops without a fold upon the quiet grass, which was dotted here and there with a humble native, dozing away the golden afternoon. The natives of the elder Carcassonne are all humble; for the core of the Cite has shrunken and decayed, and there is little life among the ruins. A few tenacious laborers, who work in the neighboring fields or in the *ville-basse*, and sundry octogenarians of both sexes, who are dying where they have lived, and contribute much to the pictorial effect, these are the principal inhabitants. The process of converting the place from an irresponsible old town into a conscious "specimen" has of course been attended with eliminations; the population has, as a general thing, been restored away. I should lose no time in saying that restoration is the great mark of the Cite. M. Viollet-le-Duc has worked his will upon it, put it into perfect order, revived the fortifications in every detail. I do not pretend to judge the performance, carried out on a scale and in a spirit which really impose themselves on the imagination. Few architects have had such a chance, and M. Viollet-le-Duc must have been the envy of the whole restoring fraternity. The image of a more crumbling Carcassonne rises in the mind, and there is no doubt that forty years ago the place was more affecting. On the other hand, as we see it to-day, it is a wonderful evocation; and if there is a great deal of new in the old, there is plenty of old in the new. The repaired crenellations, the inserted patches, of the walls of the outer circle sufficiently express this commixture. My walk brought me into full view of the Pyrenees, which, now that the sun had begun to sink and the shadows to grow long, had a wonderful violet glow. The platform at the base of the walls has a greater width on this side, and it made the scene more complete. Two or three old crones had crawled out of the Porte Narbonnaise, to examine the advancing visitor; and a very ancient peasant, lying there with his back against a tower, was tending half a dozen lean sheep. A poor man in a very old blouse, crippled and with crutches lying beside him, had been brought out and placed on a stool, where he enjoyed the afternoon as best he might. He looked so ill and so patient that I spoke to him; found that his legs were paralyzed and he was quite helpless. He had formerly been seven years in the army, and had made the campaign of Mexico with Bazaine. Born in the old Cite, he had come back there to end his days. It seemed strange, as he sat there, with those romantic walls behind him and the great picture of the Pyrenees in front, to think that he had been across the seas to the far-away new world, had made part of a famous expedition, and was now a cripple at the gate of the mediaeval city where he had played as a child. All this struck me as a great deal of history for so modest a figure, a poor little figure that could only just unclose its palm for a small silver coin.

He was not the only acquaintance I made at Carcassonne. I had not pursued my circuit of the walls much further when I encountered a person of quite another type, of whom I asked some question which had just then presented, itself, and who proved to be the very genius of the spot. He was a sociable son of the *ville-basse*, a gentleman, and, as I afterwards learned, an employee at the prefecture, a person, in short, much esteemed at Carcassonne. (I may say all this, as he will never read these pages.) He had been ill for a month, and in the company of his little dog was taking his first airing; in his own phrase he was *amoureux-fou de la Cite*, he could lose no time in coming back to it. He talked of it, indeed, as a lover, and, giving me for half an hour the advantage of his company, showed me all the points of the place. (I speak here always of the outer enceinte; you penetrate to the inner which is the specialty of Carcassonne, and the great curiosity only by application at the lodge of the regular custodian, a remarkable functionary, who, half an hour later, when I had been introduced to him by my friend the amateur, marched me over the fortifications with a tremendous accompaniment of dates and technical terms.) My companion pointed out to me in particular the traces of different periods in the structure of the walls. There is a portentous amount of history embedded in them, beginning with Romans and Visigoths; here and there are marks of old breaches, hastily repaired. We passed into the town, into that part of it not included in the citadel. It is the queerest and most fragmentary little place in the world, as everything save the fortifications is being suffered to crumble away, in order that the spirit of M. Viollet-le-Duc alone may pervade it, and it may subsist simply as a magnificent shell. As the leases of the wretched little houses fall in, the ground is cleared of them; and a mumbling old woman approached me in the course of my circuit, inviting me to condole with her on the disappearance of so many of the hovels which in the last few hundred years (since the collapse of Carcassonne as a stronghold) had attached themselves to the base of the walls, in the space between the two circles. These habitations, constructed of materials taken from the ruins, nestled there snugly enough. This intermediate space had therefore become a kind of street, which has crumbled in turn, as the fortress has grown up again. There are other streets, beside, very diminutive and vague, where you pick your way over heaps of rubbish and become conscious of unexpected faces looking at you out of windows as detached as the cherubic heads. The most definite thing in the place was the little cafe, where. the waiters, I think, must be the ghosts of the old Visigoths; the most definite, that is, after the little chateau and the little cathedral. Everything in the Cite is little; you can walk round the walls in twenty minutes. On the drawbridge of the chateau, which, with a picturesque old face, flanking towers, and a dry moat, is to-day simply a bare *caserne*, lounged half a dozen soldiers, unusually small. Nothing could be more odd than to see these objects enclosed in a receptacle which has much of the appearance of an enormous toy. The Cite and its population vaguely reminded me of an immense Noah's ark.

92

23

Carcassonne dates from the Roman occupation of Gaul. The place commanded one of the great roads into Spain, and in the fourth century Romans and Franks ousted each other from such a point of vantage. In the year 436, Theodoric, King of the Visigoths, superseded both these parties; and it is during his occupation that the inner enceinte was raised upon the ruins of the Roman fortifications. Most of the Visigoth towers that are still erect are seated upon Roman substructions which appear to have been formed hastily, probably at the moment of the Frankish invasion. The authors of these solid defenses, though occasionally disturbed, held Carcassonne and the neighboring country, in which they had established their kingdom of Septimania, till the year 713, when they were expelled by the Moors of Spain, who ushered in an unillumined period of four centuries, of which no traces remain. These facts I derived from a source no more recondite than a pamphlet by M. Viollet-le-Duc, a very luminous description of the fortifications, which you may buy from the accomplished custodian. The writer makes a jump to the year 1209, when Carcassonne, then forming part of the realm of the viscounts of Beziers and infected by the Albigensian heresy, was besieged, in the name of the Pope, by the terrible Simon de Montfort and his army of crusaders. Simon was accustomed to success, and the town succumbed in the course of a fortnight. Thirty-one years later, having passed into the hands of the King of France, it was again besieged by the young Raymond de Trincavel, the last of the viscounts of Beziers; and of this siege M. Viollet-le-Duc gives a long and minute account, which the visitor who has a head for such things may follow, with the brochure in hand, on the fortifications themselves. The young Raymond de Trincavel, baffled and repulsed, retired at the end of twenty-four days. Saint Louis and Philip the Bold, in the thirteenth century, multiplied the defenses of Carcassonne, which was one of the bulwarks of their kingdom on the Spanish quarter; and from this time forth, being regarded as impregnable, the place had nothing to fear. It was not even attacked; and when, in 1355, Edward the Black Prince marched into it, the inhabitants had opened the gates to the conqueror before whom all Languedoc was prostrate. I am not one of those who, as I said just now, have a head for such things, and having extracted these few facts had made all the use of M. Viollet-le-Duc's, pamphlet of which I was capable.

I have mentioned that my obliging friend the *amoureux-fou* handed me over to the door-keeper of the citadel. I should add that I was at first committed to the wife of this functionary, a stout peasant-woman, who took a key down from a nail, conducted me to a postern door, and ushered me into the presence of her husband. Having just begun his rounds with a party of four persons, he was not many steps in advance. I added myself perforce to this party, which was not brilliantly composed, except that two of its members were gendarmes in full toggery, who announced in the course of our tour that they had been stationed for a year at Carcassonne, and had never before had the curiosity to come up to the Cite. There was something brilliant, certainly, in that. The *gardien* was an extraordinarily typical little Frenchman, who struck me even more forcibly than the wonders of the inner enceinte; and as I am bound to assume, at whatever cost to my literary vanity, that there is not the slightest danger of his reading these remarks, I may treat him as public property. With his diminutive stature and his perpendicular spirit, his flushed face, expressive protuberant eyes, high peremptory voice, extreme volubility, lucidity, and neatness of utterance, he reminded me of the gentry who figure in the revolutions of his native land. If he was not a fierce little Jacobin, he ought to have been, for I am sure there were many men of his pattern on the Committee of Public Safety. He knew absolutely what he was about, understood the place thoroughly, and constantly reminded his audience of what he himself had done in the way of excavations and reparations. He described himself as the brother of the architect of the work actually going forward (that which has been done since the death of M. Viollet-le-Duc, I suppose he meant), and this fact was more illustrative than all the others. It reminded me, as one is reminded at every turn, of the democratic conditions of French life: a man of the people, with a wife *en bonnet*, extremely intelligent, full of special knowledge, and yet remaining essentially of the people, and showing his intelligence with a kind of ferocity, of defiance. Such a personage helps one to understand the red radicalism of France, the revolutions, the barricades, the sinister passion for theories. (I do not, of course, take upon myself to say that the individual I describe who can know nothing of the liberties I am taking with him is actually devoted to these ideals; I only mean that many such devotees must have his qualities.) In just the *nuance* that I have tried to indicate here, it is a terrible pattern of man. Permeated in a high degree by civilization, it is yet untouched by the desire which one finds in the Englishman, in proportion as he rises in the world, to approximate to the figure of the gentleman. On the other hand, a *nettete*, a faculty of exposition, such as the English gentleman is rarely either blessed or cursed with.

This brilliant, this suggestive warden of Carcassonne marched us about for an hour, haranguing, explaining, illustrating, as he went; it was a complete little lecture, such as might have been delivered at the Lowell Institute, on the manger in which a first-rate *place forte* used to be attacked and defended Our peregrinations made it very clear that Carcassone was impregnable; it is impossible to imagine, without having seen them, such refinements of immurement, such ingenuities of resistance. We passed along the battlements and *chemins de ronde*, ascended and descended towers, crawled under arches, peered out of loop-holes, lowered ourselves into dungeons, halted in all sorts of tight places, while the purpose of something or other was described to us. It was very curious, very interesting; above all, it was very pictorial, and involved perpetual peeps into the little crooked, crumbling, sunny, grassy, empty Cite. In places, as you stand upon it, the great towered and embattled enceinte produces an illusion; it looks as if it were still equipped and defended. One vivid challenge, at any rate, it flings down before you; it calls upon you to make up your mind on the matter of restoration. For myself, I have no hesitation; I prefer in every case the ruined, however ruined, to the reconstructed, however splendid. What is left is more precious than what is added: the one is history, the other is fiction; and I like the former the better of the two, it is so much more romantic. One is positive, so far as it goes; the other fills up the void with things more dead than the void itself, inasmuch as they have never had life. After that I am free to say that the restoration of Carcassonne is a splendid achievement. The little custodian dismissed us at last, after having, as usual, inducted us into the inevitable repository of photographs. These photographs are a great nuisance, all over the Midi. They are exceedingly bad, for the most part; and the worst those in the form of the hideous little *album-panorama* are thrust upon you at every turn. They are a kind of tax that you must pay; the best way is to pay to be let off. It was not to be denied that there was a relief in separating from our accomplished guide, whose manner of imparting information reminded me of the energetic process by which I have seen mineral waters bottled. All this while the afternoon had grown more lovely; the sunset had deepened, the horizon of hills grown purple; the mass of the Canigou became more delicate, yet more distinct. The day had so far faded that the interior of the little

95

cathedral was wrapped in twilight, into which the glowing windows projected something of their color. This church has high beauty and value, but I will spare the reader a presentation of details which I myself had no opportunity to master. It consists of a Romanesque nave, of the end of the eleventh century, and a Gothic choir and transepts of the beginning of the fourteenth; and, shut up in its citadel like a precious casket in a cabinet, it seems or seemed at that hour to have a sort of double sanctity. After leaving it and passing out of the two circles of walls, I treated myself, in the most infatuated manner, to another walk round the Cite. It is certainly this general impression that is most striking, the impression from outside, where the whole place detaches itself at once from the landscape. In the warm southern dusk it looked more than ever like a city in a fairy-tale. To make the thing perfect, a white young moon, in its first quarter, came out and hung just over the dark silhouette. It was hard to come away, to incommode one's self for anything so vulgar as a railway-train; I would gladly have spent the evening in revolving round the walls of Carcassonne. But I had in a measure engaged to proceed to Narborme, and there was a certain magic that name which gave me strength, Narbonne, the richest city in Roman Gaul.

24

At Narbonne I took up my abode at the house of a *serrurier mecanicien*, and was very thankful for the accommodation. It was my misfortune to arrive at this ancient city late at night, on the eve of marketday; and market-day at Narbonne is a very serious affair. The inns, on this occasion, are stuffed with wine-dealers; for the country roundabout, dedicated almost exclusively to Bacchus, has hitherto escaped the phylloxera. This deadly enemy of the grape is encamped over the Midi in a hundred places; blighted vineyards and ruined proprietors being quite the order of the day. The signs of distress are more frequent as you advance into Provence, many of the vines being laid under water, in the hope of washing the plague away. There are healthy regions still, however, and the vintners find plenty to do at Narbonne. The traffic in wine appeared to be the sole thought of the Narbonnais; every one I spoke to had something to say about the harvest of gold that bloomed under its influence. "C'est inoui, monsieur, l'argent qu'il y a dans ce pays. Des gens a qui la vente de leur vin rapporte jusqu'a 500,000 francs par an." That little speech, addressed to me by a gentleman at the inn, gives the note of these revelations. It must be said that there was little in the appearance either of the town or of its population to suggest the possession of such treasures. Narbonne is a *sale petite ville* in all the force of the term, and my first impression on arriving there was an extreme regret that I had not remained for the night at the lovely Carcassonne. My journey from that delectable spot lasted a couple of hours, and was performed in darkness, a darkness not so dense, however, but that I was able to make out, as we passed it, the great figure of Beziers, whose ancient roofs and towers, clustered on a goodly hilltop, looked as fantastic as you please. I know not what appearance Beziers may present by day; but by night it has quite the grand air. On issuing from the station at Narbonne, I found that the only vehicle in waiting was a kind of bastard tramcar, a thing shaped as if it had been meant to go upon rails; that is, equipped with small wheels, placed beneath it, and with a platform at either end, but destined to rattle over the stones like the most vulgar of omnibuses. To complete the oddity of this conveyance, it was under the supervision, not of a conductor, but of a conductress. A fair young woman, with a pouch suspended from her girdle, had command of the platform; and as soon as the car was full she jolted us into the town through clouds of the thickest dust I ever have swallowed. I have had occasion to speak of the activity of women in France, of the way they are always in the ascendant; and here was a signal example of their general utility. The young lady I have mentioned conveyed her whole company to the wretched little Hotel de France, where it is to be hoped that some of them found a lodging. For myself, I was informed that the place was crowded from cellar to attic, and that its inmates were sleeping three or four in a room. At Carcassonne I should have had a bad bed, but at Narbonne, apparently, I was to have no bed at all. I passed an hour or two of flat suspense, while fate settled the question of whether I should go on to Perpignan, return to Beziers, or still discover a modest couch at Narbonne. I shall not have suffered in vain, however, if my example serves to deter other travelers from alighting unannounced at that city on a Wednesday evening. The retreat to Beziers, not attempted in time, proved impossible, and I was assured that at Perpignan, which I should not reach till midnight, the affluence of wine-dealers was not less than at Narbonne. I interviewed every hostess in the town, and got no satisfaction but distracted shrugs. Finally, at an advanced hour, one of the servants of the Hotel de France, where I had attempted to dine, came to me in triumph to proclaim that he had secured for me a charming apartment in a *maison bourgeoise*. I took possession of it gratefully, in spite of its having an entrance like a stable, and being pervaded by an odor compared with which that of a stable would have been delicious. As I have mentioned, my landlord was a locksmith, and he had strange machines which rumbled and whirred in the rooms below my own. Nevertheless, I slept, and I dreamed of Carcassonne. It was better to do that than to dream of the Hotel de France.

I was obliged to cultivate relations with the cuisine of this establishment. Nothing could have been more *meridional*, indeed, both the dirty little inn and Narbonne at large seemed to me to have the infirmities of the south, without its usual graces. Narrow, noisy, shabby, belittered and encumbered, filled with clatter and chatter, the Hotel de France would have been described in perfection by Alphonse Daudet. For what struck me above all in it was the note of the Midi, as he has represented it, the sound of universal talk. The landlord sat at supper with sundry friends, in a kind of glass cage, with a genial indifference to arriving guests; the waiters tumbled over the loose luggage in the hall; the travelers who had been turned away leaned gloomily against door-posts; and the landlady, surrounded by confusion, unconscious of responsibility, and animated only by the spirit of conversation, bandied high-voiced compliments with the *voyageurs de commerce*. At ten o'clock in the morning there was a table d'hote for breakfast, a wonderful repast, which overflowed into every room and pervaded the whole establishment. I sat down with a hundred hungry marketers, fat, brown, greasy men, with a good deal of the rich soil of Languedoc adhering to their hands and their boots. I mention the latter articles because they almost put them on the table. It was very hot, and there were swarms of flies; the viands had the strongest odor; there was in particular a horrible mixture known as *gras-double*, a light gray, glutinous, nauseating mess, which my companions devoured in large quantities. A man opposite to me had the dirtiest fingers I ever saw; a collection of fingers which in England would have excluded him from a farmers' ordinary. The conversation was mainly bucolic; though a part of it, I remember, at the table at which I sat, consisted of a discussion as to whether or no the maidservant were *sage*, a discussion which went on under the nose of this young lady, as she carried about the dreadful *gras-double*, and to which she contributed the most convincing blushes. It was thoroughly *meridional*.

In going to Narbonne I had of course counted upon Roman remains; but when I went forth in search of them I perceived that I had hoped too fondly. There is really nothing in the place to speak of; that is, on the day of my visit there was nothing but the market, which was in complete possession. "This intricate, curious, but lifeless town," Murray calls it; yet to me it appeared overflowing with life. Its streets are mere crooked, dirty lanes, bordered with perfectly insignificant houses; but they were filled with the same clatter and chatter that I had found at the hotel. The market was held partly in the little square of the hotel de ville, a structure which a flattering wood-cut in the Guide-Joanne had given me a desire to behold. The reality was not impressive, the old color of the front having been completely restored away. Such interest as it superficially possesses it derives from a fine mediaeval tower which rises beside it, with turrets at the angles, always a picturesque thing. The rest of the market was held in another *place*, still shabbier than the first, which lies beyond the canal. The Canal du Midi flows through the town, and, spanned at this point by a small suspension-bridge, presented a certain sketchability. On the further side were the venders and chafferers, old women under awnings and big umbrellas, rickety tables piled high with fruit, white caps and brown faces, blouses, sabots, donkeys. Beneath this picture was another, a long row of washerwomen, on their knees on the edge of the canal, pounding and wringing the dirty linen of Narbonne, no great quantity, to judge by the costume of the people. Innumerable rusty men, scattered all over the place, were buying and selling wine, straddling about in pairs, in groups, with their hands in their pockets, and packed together at the doors of the cafes. They were mostly fat and brown and unshaven; they ground their teeth as they talked; they were very *meridionaux.*

The only two lions at Narbonne are the cathedral and the museum, the latter of which is quartered in the hotel de ville. The cathedral, closely shut in by houses, and with the west front undergoing repairs, is singular in two respects. It consists exclusively of a choir, which is of the end of the thirteenth century and the beginning of the next, and of great magnificence. There is absolutely nothing else. This choir, of extraordinary elevation, forms the whole church. I sat there a good while; there was no other visitor. I had taken a great dislike to poor little Narbonne, which struck me as sordid and overheated, and this place seemed to extend to me, as in the Middle Ages, the privilege of sanctuary. It is a very solemn corner. The other peculiarity of the cathedral is that, externally, it bristles with battlements, having anciently formed part of the defenses of the *archeveche*, which is beside it and which connects it with the hotel de ville. This combination of the church and the fortress is very curious, and during the Middle Ages was not without its value. The palace of the former archbishops of Narbonne (the hotel de ville of to-day forms part of it) was both an asylum and an arsenal during the hideous wars by which the Languedoc was ravaged in the thirteenth century. The whole mass of buildings is jammed together in a manner that from certain points of view makes it far from apparent which feature is which. The museum occupies several chambers at the top of the hotel de ville, and is not an imposing collection. It was closed, but I induced the portress to let me in, a silent, cadaverous person, in a black coif, like a *beguine*, who sat knitting in one of the windows while I went the rounds. The number of Roman fragments is small, and their quality is not the finest; I must add that this impression was hastily gathered. There is indeed a work of art in one of the rooms which creates a presumption in favor of the place, the portrait (rather a good one) of a citizen of Narbonne, whose name I forget, who is described as having devoted all his time and his intelligence to collecting the objects by which the. visitor is surrounded. This excellent man was a connoisseur, and the visitor is doubtless often an ignoramus.

25

"Cette, with its glistening houses white, Curves with the curving beach away To where the lighthouse beacons bright, Far in the bay."

That stanza of Matthew Arnold's, which I happened to remember, gave a certain importance to the half-hour I spent in the buffet of the station at Cette while I waited for the train to Montpellier. I had left Narbonne in the afternoon, and by the time I reached Cette the darkness had descended. I therefore missed the sight of the glistening houses, and had to console myself with that of the beacon in the bay, as well as with a *bouillon* of which I partook at the buffet aforesaid; for, since the morning, I had not ventured to return to the table d'hote at Narbonne. The Hotel Nevet, at Montpellier, which I reached an hour later, has an ancient renown all over the south of France, advertises itself, I believe, as *le plus vaste du midi*. It seemed to me the model of a good provincial inn; a big rambling, creaking establishment, with brown, labyrinthine corridors, a queer old open-air vestibule, into which the diligence, in the *bon temps*, used to penetrate, and an hospitality more expressive than that of the new caravansaries. It dates from the days when Montpellier was still accounted a fine winter residence for people with weak lungs; and this rather melancholy tradition, together with the former celebrity of the school of medicine still existing there, but from which the glory has departed, helps to account for its combination of high antiquity and vast proportions. The old hotels were usually more concentrated; but the school of medicine passed for one of the attractions of Montpellier. Long before Mentone was discovered or Colorado invented, British invalids travelled down through France in the post-chaise or the public coach to spend their winters in the wonderful place which boasted both a climate and a faculty. The air is mild, no doubt, but there are refinements of mildness which were not then suspected, and which in a more analytic age have carried the annual wave far beyond Montpellier. The place is charming, all the same; and it served the purpose of John Locke; who made a long stay there, between 1675 and 1679, and became acquainted with a noble fellow-visitor, Lord Pembroke, to whom he dedicated the famous Essay. There are places that please, without your being able to say wherefore, and Montpellier is one of the number. It has some charming views, from the great promenade of the Peyrou; but its position is not strikingly fair. Beyond this it contains a good museum and the long facades of its school, but these are its only definite treasures. Its cathedral struck me as quite the weakest I had seen, and I remember no other monument that made up for it. The place has neither the gayety of a modern nor the solemnity of an ancient town, and it is agreeable as certain women are agreeable who are neither beautiful nor clever. An Italian would remark that it is sympathetic; a German would admit that it is *gemuthlich*. I spent two days there, mostly in the rain, and even under these circumstances I carried away a kindly impression. I think the Hotel Nevet had something to do with it, and the sentiment of relief with which, in a quiet, even a luxurious, room that looked out on a garden, I reflected that I had washed my hands of Narbonne. The phylloxera has destroyed the vines in the country that surrounds Montpellier, and at that moment I was capable of rejoicing in the thought that I should not breakfast with vintners.

The gem of the place is the Musee Fabre, one of the best collections of paintings in a provincial city. Francois Fabre, a native of Montpellier, died there in 1837, after having spent a considerable part of his life in Italy, where he had collected a good many valuable pictures and some very poor ones, the latter class including several from his own hand. He was the hero of a remarkable episode, having succeeded no less a person than Vittorio Alfieri in the affections of no less a person than Louise de Stolberg, Countess of Albany, widow of no less a person than Charles Edward Stuart, the second pretender to the British crown. Surely no woman ever was associated sentimentally with three figures more diverse, a disqualified sovereign, an Italian dramatist, and a bad French painter. The productions of M. Fabre, who followed in the steps of David, bear the stamp of a cold mediocrity; there is not much to be said even for the portrait of the genial countess (her life has been written by M. Saint-Rene-Taillandier, who depicts her as delightful), which hangs in Florence, in the gallery of the Uffizzi, and makes a pendant to a likeness of Alfieri by the same author. Stendhal, in his "Memoires d'un Touriste," says that this work of art represents her as a cook who has pretty hands. I am delighted to have an opportunity of quoting Stendhal, whose two volumes of the "Memoires d'un Touriste" every traveler in France should carry in his portmanteau. I have had this opportunity more than once, for I have met him at Tours, at Nantes, at Bourges; and everywhere he is suggestive. But he has the defect that he is never pictorial, that he never by any chance makes an image, and that his style is perversely colorless, for a man so fond of contemplation. His taste is often singularly false; it is the taste of the early years of the present century, the period that produced clocks surmounted with sentimental "subjects." Stendhal does not admire these clocks, but he almost does. He admires Domenichino and Guercino, and prizes the Bolognese school of painters because they "spoke to the soul." He is a votary of the new classic, is fond of tall, squire, regular buildings, and thinks Nantes, for instance, full of the "air noble." It was a pleasure to me to reflect that five-and-forty years ago he had alighted in that city, at the very inn in which I spent a night, and which looks down on the Place Graslin and the theatre. The hotel that was the best in 1837 appears to be the best to-day. On the subject of Touraine, Stendhal is extremely refreshing; he finds the scenery meager and much overrated, and proclaims his opinion with perfect frankness. He does, however, scant justice to the banks of the Loire; his want of appreciation of the picturesque want of the sketcher's sense causes him to miss half the charm of a landscape which is nothing if not "quiet," as a painter would say, and of which the felicities reveal themselves only to waiting eyes. He even despises the Indre, the river of Madame Sand. The "Memoires d'un Touriste" are written in the character of a commercial traveler, and the author has nothing to say about Chenonceaux or Chambord, or indeed about any of the chateaux of that part of France; his system being to talk only of the large towns, where he may be supposed to find a market for his goods. It was his ambition to pass for an ironmonger. But in the large towns he is usually excellent company, though as discursive as Sterne, and strangely indifferent, for a man of imagination, to those superficial aspects of things which the poor pages now before the reader are mainly an attempt to render. It is his conviction that Alfieri, at Florence, bored the Countess of Albany terribly; and he adds that the famous Gallophobe died of jealousy of the little painter from Montpellier. The Countess of Albany left her property to Fabre; and I suppose some of the pieces in the museum of his native town used to hang in the sunny saloons of that fine old palace on the Arno which is still pointed out to the stranger in Florence as the residence of Alfieri.

The institution has had other benefactors, notably a certain M. Bruyas, who has enriched it with an extraordinary number of portraits of himself. As these, however, are by different hands, some of them distinguished, we may suppose that it was less the model than the artists to whom M. Bruyas wished to give publicity. Easily first are two large specimens of David Teniers, which are incomparable for brilliancy and a glowing perfection of execution. I have a weakness for this singular genius, who combined the delicate with the groveling, and I have rarely seen richer examples. Scarcely less valuable is a Gerard Dow which hangs near them, though it must rank lower as having kept less of its freshness. This Gerard Dow did me good; for a master is a master, whatever he may paint. It represents a woman paring carrots, while a boy before her exhibits a mouse-trap in which he has caught a frightened victim. The good-wife has spread a cloth on the top of a big barrel which serves her as a table, and on this brown, greasy napkin, of which the texture is wonderfully rendered, lie the raw vegetables she is preparing for domestic consumption. Beside the barrel is a large caldron lined with copper, with a rim of brass. The way these things are painted brings tears to the eyes; but they give the measure of the Musee Fabre, where two specimens of Teniers and a Gerard Dow are the jewels. The Italian pictures are of small value; but there is a work by Sir Joshua Reynolds, said to be the only one in France, an infant Samuel in prayer, apparently a repetition of the picture in England which inspired the little plaster image, disseminated in Protestant lands, that we used to admire in our childhood. Sir Joshua, somehow, was an eminently Protestant painter; no one can forget that, who in the National Gallery in London has looked at the picture in which he represents several young ladies as nymphs, voluminously draped, hanging garlands over a statue, a picture suffused indefinably with the Anglican spirit, and exasperating to a member of one of the Latin races. It is an odd chance, therefore, that has led him into that part of France where Protestants have been least *bien vus.* This is the country of the dragonnades of Louis XIV. and of the pastors of the desert. From the garden of the Peyrou, at Montpellier, you may see the hills of the Cevennes, to which they of the religion fled for safety, and out of which they were hunted and harried.

I have only to add, in regard to the Musee Fabre, that it contains the portrait of its founder, a little, pursy, fat-faced, elderly man, whose countenance contains few indications of the power that makes distinguished victims. He is, however, just such a personage as the mind's eye sees walking on the terrace of the Peyrou of an October afternoon in the early years of the century; a plump figure in a chocolate-colored coat and a *culotte* that exhibits a good leg, a culotte provided with a watch-fob from which a heavy seal is suspended. This Peyrou (to come to it at last) is a wonderful place, especially to be found in a little provincial city. France is certainly the country of towns that aim at completeness; more than in other lands, they contain stately features as a matter of course. We should never have ceased to hear about the Peyrou, if fortune had placed it at a Shrewsbury or a Buffalo. It is true that the place enjoys a certain celebrity at home, which it amply deserves, moreover; for nothing could be more impressive and monumental. It consists of an "elevated platform," as Murray says, an immense terrace, laid out, in the highest part of the town, as a garden, and commanding in all directions a view which in clear weather must be of the finest. I strolled there in the intervals of showers, and saw only the nearer beauties, a great pompous arch of triumph in honor of Louis XIV. (which is not, properly speaking, in the garden, but faces it, straddling across the *place* by which you approach it from the town), an equestrian statue of that monarch set aloft in the middle of the terrace, and a very exalted and complicated fountain, which forms a background to the picture. This fountain gushes from a kind of hydraulic temple, or *chateau d'eau,* to which you ascend by broad flights of steps, and which is fed by a splendid aqueduct, stretched in the most ornamental and unexpected manner across the neighboring valley. All this work dates from the middle of the last century. The combination of features the triumphal arch, or gate; the wide, fair terrace, with its beautiful view; the statue of the grand monarch; the big architectural fountain, which would not surprise one at Rome, but goes surprise one at Montpellier; and to complete the effect, the extraordinary aqueduct, charmingly fore-shortened, all this is worthy of a capital, of a little court-city. The whole place, with its repeated steps, its balustrades, its massive and plentiful stone-work, is full of the air of the last century, *sent bien son dix-huitieme siecle*, none the less so, I am afraid, that, as I read in my faithful Murray, after the revocation of the Edict of Nantes, the block, the stake, the wheel, had been erected here for the benefit of the desperate Camisards.

26

It was a pleasure to feel one's self in Provence again, the land where the silver-gray earth is impregnated with the light of the sky. To celebrate the event, as soon as I arrived at Nimes I engaged a caleche to convey me to the Pont du Gard. The day was yet young, and it was perfectly fair; it appeared well, for a longish drive, to take advantage, without delay, of such security. After I had left the town I became more intimate with that Provencal charm which I had already enjoyed from the window of the train, and which glowed in the sweet sunshine and the white rocks, and lurked in the smoke-puffs of the little olives. The olive-trees in Provence are half the landscape. They are neither so tall, so stout, nor so richly contorted as I have seen them beyond the Alps; but this mild colorless bloom seems the very texture of the country. The road from Nimes, for a distance of fifteen miles, is superb; broad enough for an army, and as white and firm as a dinner-table. It stretches away over undulations which suggest a kind of harmony; and in the curves it makes through the wide, free country, where there is never a hedge or a wall, and the detail is always exquisite, there is something majestic, almost processional. Some twenty minutes before I reached the little inn that marks the termination of the drive, my vehicle met with an accident which just missed being serious, and which engaged the attention of a gentleman, who, followed by his groom and mounted on a strikingly handsome horse happened to ride up at the moment. This young man, who, with his good looks and charming manner, might have stepped out of a novel of Octave Feuillet, gave me some very intelligent advice in reference to one of my horses that had been injured, and was so good as to accompany me to the inn, with the resources of which he was acquainted, to see that his recommendations were carried out. The result of our interview was that he invited me to come and look at a small but ancient chateau in the neighborhood, which he had the happiness not the greatest in the world, he intimated to inhabit, and at which I engaged to present myself after I should have spent an hour at the Pont du Gard. For the moment, when we separated, I gave all my attention to that great structure. You are very near it before you see it; the ravine it spans suddenly opens and exhibits the picture. The scene at this point grows extremely beautiful. The ravine is the valley of the Gardon, which the road from Nimes has followed some time without taking account of it, but which, exactly at the right distance from the aqueduct, deepens and expands, and puts on those characteristics which are best suited to give it effect. The gorge becomes romantic, still, and solitary, and, with its white rocks and wild shrubbery, hangs over the clear, colored river, in whose slow course there is here and there a deeper pool. Over the valley, from side to side, and ever so high in the air, stretch the three tiers of the tremendous bridge. They are unspeakably imposing, and nothing could well be more Roman. The hugeness, the solidity, the unexpectedness, the monumental rectitude of the whole thing leave you nothing to say at the time and make you stand gazing. You simply feel that it is noble and perfect, that it has the quality of greatness. A road, branching from the highway, descends to the level of the river and passes under one of the arches. This road has a wide margin of grass and loose stones, which slopes upward into the bank of the ravine. You may sit here as long as you please, staring up at the light, strong piers; the spot is extremely natural, though two or three stone benches have been erected on it. I remained there an hour and got a cornplete impression; the place was perfectly soundless, and for the time, at least, lonely; the splendid afternoon had begun to fade, and there was a fascination in the object I had come to see. It came to pass that at the same time I discovered in it a certain stupidity, a vague brutality. That element is rarely absent from great Roman work, which is wanting in the nice adaptation of the means to the end. The means are always exaggerated; the end is so much more than attained. The Roman rigidity was apt to overshoot the mark, and I suppose a race which could do nothing small is as defective as a race that can do nothing great. Of this Roman rigidity the Pont du Gard is an admirable example. It would be a great injustice, however, not to insist upon its beauty, a kind of manly beauty, that of an object constructed not to please but to serve, and impressive simply from the scale on which it carries out this intention. The number of arches in each tier is different; they are smaller and more numerous as they ascend. The preservation of the thing is extraordinary; nothing has crumbled or collapsed; every feature remains; and the huge blocks of stone, of a brownish-yellow, (as if they had been baked by the Provencal sun for eighteen centuries), pile themselves, without mortar or cement, as evenly as the day they were laid together. All this to carry the water of a couple of springs to a little provincial city! The conduit on the top has retained its shape and traces of the cement with which it was lined. When the vague twilight began to gather, the lonely valley seemed to fill itself with the shadow of the Roman name, as if the mighty empire were still as erect as the supports of the aqueduct; and it was open to a solitary tourist, sitting there sentimental, to believe that no people has ever been, or will ever be, as great as that, measured, as we measure the greatness of an individual, by the push they gave to what they undertook. The Pont du Gard is one of the three or four deepest impressions they have left; it speaks of them in a manner with which they might have been satisfied.

I feel as if it were scarcely discreet to indicate the whereabouts of the chateau of the obliging young man I had met on the way from Nimes; I must content myself with saying that it nestled in an enchanting valley, *dans le fond*, as they say in France, and that I took my course thither on foot, after leaving the Pont du Gard. I find it noted in my journal as "an adorable little corner." The principal feature of the place is a couple of very ancient towers, brownish-yellow in hue, and mantled in scarlet Virginia-creeper. One of these towers, reputed to be of Saracenic origin, is isolated, and is only the more effective; the other is incorporated in the house, which is delightfully fragmentary and irregular. It had got to be late by this time, and the lonely *castel* looked crepuscular and mysterious. An old housekeeper was sent for, who showed me the rambling interior; and then the young man took me into a dim old drawing-room, which had no less than four chimney-pieces, all unlighted, and gave me a refection of fruit and sweet wine. When I praised the wine and asked him what it was, he said simply, "C'est du vin de ma mere!" Throughout my little journey I had never yet felt myself so far from Paris; and this was a sensation I enjoyed more than my host, who was an involuntary exile, consoling himself with laying out a *manege*, which he showed me as I walked away. His civility was great, and I was greatly touched by it. On my way back to the little inn where I had left my vehicle, I passed the Pont du Gard, and took another look at it. Its great arches made windows for the evening sky, and the rocky ravine, with its dusky cedars and shining river, was lonelier than before. At the inn I swallowed, or tried to swallow, a glass of horrible wine with my coachman; after which, with my reconstructed team, I drove back to Nimes in the moonlight. It only added a more solitary whiteness to the constant sheen of the Provencal landscape.

27

The weather the next day was equally fair, so that it seemed an imprudence not to make sure of AiguesMortes. Nimes itself could wait; at a pinch, I could attend to Nimes in the rain. It was my belief that Aigues-Mortes was a little gem, and it is natural to desire that gems should have an opportunity to sparkle. This is an excursion of but a few hours, and there is a little friendly, familiar, dawdling train that will convey you, in time for a noonday breakfast, to the small dead town where the blessed Saint-Louis twice embarked for the crusades. You may get back to Nimes for dinner; the run or rather the walk, for the train doesn't run is of about an hour. I found the little journey charming, and looked out of the carriage window, on my right, at the distant Cevennes, covered with tones of amber and blue, and, all around, at vineyards red with the touch of October. The grapes were gone, but the plants had a color of their own. Within a certain distance of Aigues-Mortes they give place to wide salt-marshes, traversed by two canals; and over this expanse the train rumbles slowly upon a narrow causeway, failing for some time, though you know you are near the object of your curiosity, to bring you to sight of anything but the horizon. Suddenly it appears, the towered and embattled mass, lying so low that the crest of its defenses seems to rise straight out of the ground; and it is not till the train stops, close before them, that you are able to take the full measure of its walls.

Aigues-Mortes stands on the edge of a wide *etang*, or shallow inlet of the sea, the further side of which is divided by a narrow band of coast from the Gulf of Lyons. Next after Carcassonne, to which it forms an admirable *pendant*, it is the most perfect thing of the kind in France. It has a rival in the person of Avignon, but the ramparts of Avignon are much less effective. Like Carcassonne, it is completely surrounded with its old fortifications; and if they are far simpler in character (there is but one circle), they are quite as well preserved. The moat has been filled up, and the site of the town might be figured by a billiard-table without pockets. On this absolute level, covered with coarse grass, Aigues-Mortes presents quite the appearance of the walled town that a school-boy draws upon his slate, or that we see in the background of early Flemish pictures, a simple parallelogram, of a contour almost absurdly bare, broken at intervals by angular towers and square holes. Such, literally speaking, is this delightful little city, which needs to be seen to tell its full story. It is extraordinarily pictorial, and if it is a very small sister of Carcassonne, it has at least the essential features of the family. Indeed, it is even more like an image and less like a reality than Carcassonne; for by position and prospect it seems even more detached from the life of the present day. It is true that Aigues-Mortes does a little business; it sees certain bags of salt piled into barges which stand in a canal beside it, and which carry their cargo into actual places. But nothing could well be more drowsy and desultory than this industry as I saw it practiced, with the aid of two or three brown peasants and under the eye of a solitary douanier, who strolled on the little quay beneath the western wall. "C'est bien plaisant, c'est bien paisible," said this worthy man, with whom I had some conversation; and pleasant and peaceful is the place indeed, though the former of these epithets may suggest an element of gayety in which Aigues-Mortes is deficient. The sand, the salt, the dull sea-view, surround it with a bright, quiet melancholy. There are fifteen towers and nine gates, five of which are on the southern side, overlooking the water. I walked all round the place three times (it doesn't take long), but lingered most under the southern wall, where the afternoon light slept in the dreamiest, sweetest way. I sat down on an old stone, and looked away to the desolate saltmarshes and the still, shining surface of the *etang*, and, as I did so, reflected that this was a queer little out-of-the-world corner to have been chosen, in the great dominions of either monarch, for that pompous interview which took place, in 1538, between Francis I. and Charles V. It was also not easy to perceive how Louis IX., when in 1248 and 1270 he started for the Holy Land, set his army afloat in such very undeveloped channels. An hour later I purchased in the town a little pamphlet by M. Marius Topin, who undertakes to explain this latter anomaly, and to show that there is water enough in the port, as we may call it by courtesy, to have sustained a fleet of crusaders. I was unable to trace the channel that he points out, but was glad to believe that, as he contends, the sea has not retreated from the town since the thirteenth century. It was comfortable to think that things are not so changed as that. M. Topin indicates that the other French ports of the Mediterranean were not then *disponsibles*, and that Aigues-Mortes was the most eligible spot for an embarkation.

Behind the straight walls and the quiet gates the little town has not crumbled, like the Cite of Carcassonne. It can hardly be said to be alive; but if it is dead it has been very neatly embalmed. The hand of the restorer rests on it constantly; but this artist has not, as at Carcassonne, had miracles to accomplish. The interior is very still and empty, with small stony, whitewashed streets, tenanted by a stray dog, a stray cat, a stray old woman. In the middle is a little *place*, with two or three cafes decorated by wide awnings, a little *place* of which the principal feature is a very bad bronze statue of Saint Louis by Pradier. It is almost as bad as the breakfast I had at the inn that bears the name of that pious monarch. You may walk round the enceinte of Aigues-Mortes, both outside and in; but you may not, as at Carcassonne, make a portion of this circuit on the *chemin de ronde*, the little projecting footway attached to the inner face of the battlements. This footway, wide enough only for a single pedestrian, is in the best order, and near each of the gates a flight of steps leads up to it; but a locked gate, at the top of the steps, makes access impossible, or at least unlawful. Aigues-Mortes, however, has its citadel, an immense tower, larger than any of the others, a little detached, and standing at the northwest angle of the town. I called upon the *casernier*, the custodian of the walls, and in his absence I was conducted through this big Tour de Constance by his wife, a very mild, meek woman, yellow with the traces of fever and ague, a scourge which, as might be expected in a town whose name denotes "dead waters," enters freely at the nine gates. The Tour de Constance is of extraordinary girth and solidity, divided into three superposed circular chambers, with very fine vaults, which are lighted by embrasures of prodigious depth, converging to windows little larger than loopholes. The place served for years as a prison to many of the Protestants of the south whom the revocation of the Edict of Nantes had exposed to atrocious penalties, and the annals of these dreadful chambers during the first half of the last century were written in tears and blood. Some of the recorded cases of long confinement there make one marvel afresh at what man has inflicted and endured. In a country in which a policy of extermination was to be put into practice this horrible tower was an obvious resource. From the battlements at the top, which is surmounted by an old disused light-house, you see the little compact rectangular town, which looks hardly bigger than a garden-patch, mapped out beneath you, and follow the plain configuration of its defenses. You take possession of it, and you feel that you will remember it always.

28

After this I was free to look about me at Nimes, and I did so with such attention as the place appeared to require. At the risk of seeming too easily and too frequently disappointed, I will say that it required rather less than I had been prepared to give. It is a town of three or four fine features, rather than a town with, as I may say, a general figure. In general, Nimes is poor; its only treasures are its Roman remains, which are of the first order. The new French fashions prevail in many of its streets; the old houses are paltry, and the good houses are new; while beside my hotel rose a big spick-and-span church, which had the oddest air of having been intended for Brooklyn or Cleveland. It is true that this church looked out on a square completely French, a square of a fine modern disposition, flanked on one side by a classical *palais de justice* embellished with trees and parapets, and occupied in the centre with a group of allegorical statues, such as one encounters only in the cities of France, the chief of these being a colossal figure by Pradier, representing Nimes. An English, an American, town which should have such a monument, such a square, as this, would be a place of great pretensions; but like so many little *villes de province* in the country of which I write, Nimes is easily ornamental. What nobler ornament can there be than the Roman baths at the foot of Mont Cavalier, and the delightful old garden that surrounds them? All that quarter of Nimes has every reason to be proud of itself; it has been revealed to the world at large by copious photography. A clear, abundant stream gushes from the foot of a high hill (covered with trees and laid out in paths), and is distributed into basins which sufficiently refer themselves to the period that gave them birth, the

111

period that has left its stamp on that pompous Peyrou which we admired at Montpellier. Here are the same terraces and steps and balustrades, and a system of water-works less impressive, perhaps, but very ingenious and charming. The whole place is a mixture of old Rome and of the French eighteenth century; for the remains of the antique baths are in a measure incorporated in the modern fountains. In a corner of this umbrageous precinct stands a small Roman ruin, which is known as a temple of Diana, but was more apparently a *nymphaeum*, and appears to have had a graceful connection with the adjacent baths. I learn from Murray that this little temple, of the period of Augustus, "was reduced to its present state of ruin in 1577;" the moment at which the townspeople, threatened with a siege by the troops of the crown, partly demolished it, lest it should serve as a cover to the enemy. The remains are very fragmentary, but they serve to show that the place was lovely. I spent half an hour in it on a perfect Sunday morning (it is enclosed by a high *grille*, carefully tended, and has a warden of its own), and with the help of my imagination tried to reconstruct a little the aspect of things in the Gallo-Roman days. I do wrong, perhaps, to say that I *tried*, from a flight so deliberate I should have shrunk. But there was a certain contagion of antiquity in the air; and among the ruins of baths and temples, in the very spot where the aqueduct that crosses the Gardon in the wondrous manner I had seen discharged itself, the picture of a splendid paganism seemed vaguely to glow. Roman baths, Roman baths; those words alone were a scene. Everything was changed: I was strolling in a *jardin francais*; the bosky slope of the Mont Cavalier (a very modest mountain), hanging over the place, is crowned with a shapeless tower, which is as likely to be of mediaeval as of antique origin; and yet, as I leaned on the parapet of one of the fountains, where a flight of curved steps (a hemicycle, as the French say) descended into a basin full of dark, cool recesses, where the slabs of the Roman foundations gleam through the clear green water, as in this attitude I surrendered myself to contemplation and reverie, it seemed to me that I touched for a moment the ancient world. Such moments are illuminating, and the light of this one mingles, in my memory, with the dusky greenness of the Jardin de la Fontaine.

The fountain proper the source of all these distributed waters is the prettiest thing in the world, a reduced copy of Vaucluse. It gushes up at the foot of the Mont Cavalier, at a point where that eminence rises with a certain cliff-like effect, and, like other springs in the same circumstances, appears to issue from the rock with a sort of quivering stillness. I trudged up the Mont Cavalier, it is a matter of five minutes, and having committed this cockneyism enhanced it presently by another. I ascended the stupid Tour Magne, the mysterious structure I mentioned a moment ago. The only feature of this dateless tube, except the inevitable collection of photographs to which you are introduced by the door-keeper, is the view you enjoy from its summit. This view is, of course, remarkably fine, but I am ashamed to say I have not the smallest recollection of it; for while I looked into the brilliant spaces of the air I seemed still to see only what I saw in the depths of the Roman baths, the image, disastrously confused and vague, of a vanished world. This world, however, has left at Nimes a far more considerable memento than a few old stones covered with water-moss. The Roman arena is the rival of those of Verona and of Arles; at a respectful distance it emulates the Colosseum. It is a small Colosseum, if I may be allowed the expression, and is in a much better preservation than the great circus at Rome. This is especially true of the external walls, with their arches, pillars, cornices. I must add that one should not speak of preservation, in regard to the arena at Nimes, without speaking also of repair. After the great ruin ceased to be despoiled, it began to be protected, and most of its wounds have been dressed with new material. These matters concern the archaeologist; and I felt here, as I felt afterwards at Arles, that one of the profane, in the presence of such a monument, can only admire and hold his tongue. The great impression, on the whole, is an impression of wonder that so much should have survived. What remains at Nimes, after all dilapidation is estimated, is astounding. I spent an hour in the Arenes on that same sweet Sunday morning, as I came back from the Roman baths, and saw that the corridors, the vaults, the staircases, the external casing, are still virtually there. Many of these parts are wanting in the Colosseum, whose sublimity of size, however, can afford to dispense with detail. The seats at Nimes, like those at Verona, have been largely renewed; not that this mattered much, as I lounged on the cool surface of one of them, and admired the mighty concavity of the place and the elliptical skyline, broken by uneven blocks and forming the rim of the monstrous cup, a cup that had been filled with horrors. And yet I made my reflections; I said to myself that though a Roman arena is one of the most impressive of the works of man, it has a touch of that same stupidity which I ventured to discover in the Pont du Gard. It is brutal; it is monotonous; it is not at all exquisite. The Arenes at Nimes were arranged for a bull-fight, a form of recreation that, as I was informed, is much *dans les habitudes Nimoises*, and very common throughout Provence, where (still according to my information) it is the usual pastime of a Sunday

113

afternoon. At Arles and Nimes it has a characteristic setting, but in the villages the patrons of the game make a circle of carts and barrels, on which the spectators perch themselves. I was surprised at the prevalence, in mild Provence, of the Iberian vice, and hardly know whether it makes the custom more respectable that at Nimes and Arles the thing is shabbily and imperfectly done. The bulls are rarely killed, and indeed often are bulls only in the Irish sense of the term, being domestic and motherly cows. Such an entertainment of course does not supply to the arena that element of the exquisite which I spoke of as wanting. The exquisite at Nimes is mainly represented by the famous Maison Carree. The first impression you receive from this delicate little building, as you stand before it, is that you have already seen it many times. Photographs, engravings, models, medals, have placed it definitely in your eye, so that from the sentiment with which you regard it curiosity and surprise are almost completely, and perhaps deplorably, absent. Admiration remains, however, admiration of a familiar and even slightly patronizing kind. The Maison Carree does not overwhelm you; you can conceive it. It is not one of the great sensations of the antique art; but it is perfectly felicitous, and, in spite of having been put to all sorts of incongruous uses, marvelously preserved. Its slender columns, its delicate proportions, its charming compactness, seemed to bring one nearer to the century that built it than the great superpositions of arenas and bridges, and give it the interest that vibrates from one age to another when the note of taste is struck. If anything were needed to make this little toy-temple a happy production, the service would be rendered by the second-rate boulevard that conducts to it, adorned with inferior cafes and tobacco-shops. Here, in a respectable recess, surrounded by vulgar habitations, and with the theatre, of a classic pretension, opposite, stands the small "square house," so called because it is much longer than it is broad. I saw it first in the evening, in the vague moonlight, which made it look as if it were cast in bronze. Stendhal says, justly, that it has the shape of a playing-card, and he expresses his admiration for it by the singular wish that an "exact copy" of it should be erected in Paris. He even goes so far as to say that in the year 1880 this tribute will have been rendered to its charms; nothing would be more simple, to his mind, than to "have" in that city "le Pantheon de Rome, quelques temples de Grece." Stendhal found it amusing to write in the character of a *commis-voyageur*, and sometimes it occurs to his reader that he really was one.

29

On my way from Nimes to Arles, I spent three hours at Tarascon; chiefly for the love of Alphonse Daudet, who has written nothing more genial than "Les Aventures Prodigieuses de Taitarin," and the story of the "siege" of the bright, dead little town (a mythic siege by the Prussians) in the "Conies du Lundi." In the introduction which, for the new edition of his works, he has lately supplied to "Tartarin," the author of this extravagant but kindly satire gives some account of the displeasure with which he has been visited by the ticklish Tarasconnais. Daudet relates that in his attempt to shed a humorous light upon some of the more erratic phases of the Provencal character, he selected Tarascon at a venture; not because the temperament of its natives is more vainglorious than that of their neighbors, or their rebellion against the "despotism of fact" more marked, but simply because he had to name a particular Provencal city. Tartarin is a hunter of lions and charmer of women, a true "*produit du midi*," as Daudet says, who has the most fantastic and fabulous adventures. He is a minimized Don Quixote, with much less dignity, but with equal good faith; and the story of his exploits is a little masterpiece of the light comical. The Tarasconnais, however, declined to take the joke, and opened the vials of their wrath upon the mocking child of Nimes, who would have been better employed, they doubtless thought, in showing up the infirmities of his own family. I am bound to add that when I passed through Tarascon they did not appear to be in the least out of humor. Nothing could have been brighter, softer, more suggestive of amiable indifference, than the picture it presented to my mind. It lies quietly bcside the Rhone, looking across at Beaucaire, which seems very distant and independent, and tacitly consenting to let the castle of the good King Rene of Anjou, which projects very boldly into the river, pass for its most interesting feature. The other features are, primarily, a sort of vivid sleepiness in the aspect of the place, as if the September noon (it had lingered on into October) lasted longer there than elsewhere; certain low arcades, which make the streets look gray and exhibit empty vistas; and a very curious and beautiful walk beside the Rhone, denominated the Chaussee, a long and narrow causeway, densely shaded by two rows of magnificent old trees, planted in its embankment, and rendered doubly effective, at the moment I passed over it, by a little train of collegians, who had been taken out for mild exercise by a pair of young priests. Lastly, one may say that a striking element of Tarascon, as of any town that lies on the Rhone, is simply the Rhone itself: the big brown flood, of uncertain temper, which has never taken time to forget that it is a child of the mountain and the glacier, and that such an origin carries with it great privileges. Later, at Avignon, I observed it in the exercise of these privileges, chief among which was that of frightening the good people of the old papal city half out of their wits.

The chateau of King Rene serves to-day as the prison of a district, and the traveler who wishes to look into it must obtain his permission at the *Mairie of Tarascon*. If he have had a certain experience of French manners, his application will be accompanied with the forms of a considerable obsequiosity, and in this case his request will be granted as civilly as it has been made. The castle has more of the air of a severely feudal fortress than I should suppose the period of its construction (the first half of the fifteenth century) would have warranted; being tremendously bare and perpendicular, and constructed for comfort only in the sense that it was arranged for defense. It is a square and simple mass, composed of small yellow stones, and perched on a pedestal of rock which easily commands the river. The building has the usual circular towers at the corners, and a heavy cornice at the top, and immense stretches of sun-scorched wall, relieved at wide intervals by small windows, heavily cross-barred. It has, above all, an extreme steepness of aspect; I cannot express it otherwise. The walls are as sheer and inhospitable as precipices. The castle has kept its large moat, which is now a hollow filled with wild plants. To this tall fortress the good Rene retired in the middle of the fifteenth century, finding it apparently the most substantial thing left him in a dominion which had included Naples and Sicily, Lorraine and Anjou. He had been a much-tried monarch and the sport of a various fortune, fighting half his life for thrones he didn't care for, and exalted only to be quickly cast down. Provence was the country of his affection, and the memory of his troubles did not prevent him from holding a joyous court at Tarascon and at Aix. He finished the castle at Tarascon, which had been begun earlier in the century, finished it, I suppose, for consistency's sake, in the manner in which it had originally been designed rather than in accordance with the artistic tastes that formed the consolation of his old age. He was a painter, a writer, a dramatist, a modern dilettante, addicted to private theatricals. There is something very attractive in the image that he has imprinted on the page of history. He was both clever and kind, and many reverses and much suffering had not embittered him nor quenched his faculty of enjoyment. He was fond of his sweet Provence, and his sweet Provence has been grateful; it has woven a light tissue of legend around the memory of the good King Rene.

I strolled over his dusky habitation it must have taken all his good-humor to light it up at the heels of the custodian, who showed me the usual number of castle-properties: a deep, well-like court; a collection of winding staircases and vaulted chambers, the embrasures of whose windows and the recesses of whose doorways reveal a tremendous thickness of wall. These things constitute the general identity of old castles; and when one has wandered through a good many, with due discretion of step and protrusion of head, one ceases very much to distinguish and remember, and contents one's self with consigning them to the honorable limbo of the romantic. I must add that this reflection did not the least deter me from crossing the bridge which connects Tarascon with Beaucaire, in order to examine the old fortress whose ruins adorn the latter city. It stands on a foundation of rock much higher than that of Tarascon, and looks over with a melancholy expression at its better-conditioned brother. Its position is magnificent, and its outline very gallant. I was well rewarded for my pilgrimage; for if the castle of Beaucaire is only a fragment, the whole place, with its position and its views, is an ineffaceable picture. It was the stronghold of the Montmorencys, and its last tenant was that rash Duke Francois, whom Richelieu, seizing every occasion to trample on a great noble, caused to be beheaded at Toulouse, where we saw, in the Capitol, the butcher's knife with which the cardinal pruned the crown of France of its thorns. The castle, after the death of this victim, was virtually demolished. Its site, which Nature to-day has taken again to herself, has an extraordinary charm. The mass of rock that it formerly covered rises high above the town, and is as precipitous as the side of the Rhone. A tall rusty iron gate admits you from a quiet corner of Beaucaire to a wild tangled garden, covering the side of the hill, for the whole place forms the public promenade of the townsfolk, a garden without flowers, with little steep, rough paths that wind under a plantation of small, scrubby stone-pines. Above this is the grassy platform of the castle, enclosed on one side only (toward the river) by a large fragment of wall and a very massive dungeon. There are benches placed in the lee of the wall, and others on the edge of the platform, where one may enjoy a view, beyond the river, of certain peeled and scorched undulations. A sweet desolation, an everlasting peace, seemed to hang in the air. A very old man (a fragment, like the castle itself) emerged from some crumbling corner to do me the honors, a very gentle, obsequious, tottering, toothless, grateful old man. He beguiled me into an ascent of the solitary tower, from which you may look down on the big sallow river and glance at diminished Tarascon, and the barefaced, bald-headed hills behind it. It may appear that I insist too much upon the nudity of the Provencal horizon, too much, considering that I have spoken of the prospect from the heights of Beaucaire as lovely. But it is an exquisite bareness; it seems to exist for the purpose of allowing one to follow the delicate lines of the hills, and touch with the eyes, as it were, the smallest inflections of the landscape. It makes the whole thing seem wonderfully bright and pure.

Beaucaire used to be the scene of a famous fair, the great fair of the south of France. It has gone the way of most fairs, even in France, where these delightful exhibitions hold their own much better than might be supposed. It is still held in the month of July; but the bourgeoises of Tarascon send to the Magasin du Louvre for their smart dresses, and the principal glory of the scene is its long tradition. Even now, however, it ought to be the prettiest of all fairs, for it takes place in a charming wood which lies just beneath the castle, beside the Rhone. The booths, the barracks, the platforms of the mountebanks, the bright-colored crowd, diffused through this midsummer shade, and spotted here and there with the rich Provencal sunshine must be of the most pictorial effect. It is highly probable, too, that it offers a large collection of pretty faces; for even in the few hours that I spent at Tarascon I discovered symptoms of the purity of feature for which the women of the *pays d'Arles* are renowned. The Arlesian head-dress, was visible in the streets; and this delightful coiffure is so associated with a charming facial oval, a dark mild eye, a straight Greek nose, and a mouth worthy of all the rest, that it conveys a presumption of beauty which gives the wearer time either to escape or to please you. I have read somewhere, however, that Tarascon is supposed to produce handsome men, as Arles is known to deal in handsome women. It may be that I should have found the Tarasconnais very fine fellows, if I had encountered enough specimens to justify an induction. But there were very few males in the streets, and the place presented no appearance of activity. Here and there the black coif of an old woman or of a young girl was framed by a low doorway; but for the rest, as I have said, Tarascon was mostly involved in a siesta. There was not a creature in the little church of Saint Martha, which I made a point of visiting before I returned to the station, and which, with its fine Romanesque sideportal and its pointed and crocketed Gothic spire, is as curious as it need be, in view of its tradition. It stands in a quiet corner where the grass grows between the small cobble-stones, and you pass beneath a deep archway to reach it. The tradition relates that Saint Martha tamed with her own hands, and attached to her girdle, a dreadful dragon, who was known as the Tarasque, and is reported to have given his name to the city on whose site (amid the rocks which form the base of the chateau) he had his cavern. The dragon, perhaps, is the symbol of a ravening paganism, dispelled by the eloquence of a sweet evangelist. The bones of the interesting saint, at all events, were found, in the eleventh century, in a cave beneath the spot on which her altar now stands. I know not what had become of the bones of the dragon.

30

There are two shabby old inns at Arles, which compete closely for your custom. I mean by this that if you elect to go to the Hotel du Forum, the Hotel du Nord, which is placed exactly beside it (at a right angle) watches your arrival with ill-concealed disapproval; and if you take the chances of its neighbor, the Hotel du Forum seems to glare at you invidiously from all its windows and doors. I forget which of these establishments I selected; whichever it was, I wished very much that, it had been the other. The two stand together on the Place des Hommes, a little public square of Arles, which somehow quite misses its effect. As a city, indeed, Arles quite misses its effect in every way; and if it is a charming place, as I think it is, I can hardly tell the reason why. The straight-nosed Arlesiennes account for it in some degree; and the remainder may be charged to the ruins of the arena and the theatre. Beyond this, I remember with affection the ill-proportioned little Place des Hommes; not at all monumental, and given over to puddles and to shabby cafes. I recall with tenderness the tortuous and featureless streets, which looked like the streets of a village, and were paved with villanous little sharp stones, making all exercise penitential. Consecrated by association is even a tiresome walk that I took the evening I arrived, with the purpose of obtaining a view of the Rhone. I had been to Arles before, years ago, and it seemed to me that I remembered finding on the banks of the stream some sort of picture. I think that on the evening of which I speak there was a watery moon, which it seemed to me would light up the past as well as the present. But I found no picture, and I scarcely found the Rhone at all. I lost my way, and there was not a creature in the streets to whom I could appeal. Nothing could be more provincial than the situation of Arles at ten o'clock at night. At last I arrived at a kind of embankment, where I could see the great mud-colored stream slipping along in the soundless darkness. It had come on to rain, I know not what had happened to the moon, and the whole place was anything but gay. It was not what I had looked for; what I had looked for was in the irrecoverable past. I groped my way back to the inn over the infernal *cailloux*, feeling like a discomfited Dogberry. I remember now that this hotel was the one (whichever that may be) which has the fragment of a Gallo-Roman portico inserted into one of its angles. I had chosen it for the sake of this exceptional ornament. It was damp and dark, and the floors felt gritty to the feet; it was an establishment at which the dreadful *gras-double* might have appeared at the table d'hote, as it had done at Narbonne. Nevertheless, I was glad to get back to it; and nevertheless, too, and this is the moral of my simple anecdote, my pointless little walk (I don't speak of the pavement) suffuses itself, as I look back upon it, with a romantic tone. And in relation to the inn, I suppose I had better mention that I am well aware of the inconsistency of a person who dislikes the modern caravansary, and yet grumbles when he finds a hotel of the superannuated sort. One ought to choose, it would seem, and make the best of either alternative. The two old taverns at Arles are quite unimproved; such as they must have been in the infancy of the modern world, when Stendhal passed that way, and the lumbering diligence deposited him in the Place des Hommes, such in every detail they are to-day. *Vieilles auberges de France*, one ought to enjoy their gritty floors and greasy window-panes. Let it be put on record, therefore, that I have been, I won't say less comfortable, but at least less happy, at better inns.

To be really historic, I should have mentioned that before going to look for the Rhone I had spent part of the evening on the opposite side of the little place, and that I indulged in this recreation for two definite reasons. One of these was that I had an opportunity of conversing at a cafe with an attractive young Englishman, whom I had met in the afternoon at Tarascon, and more remotely, in other years, in London; the other was that there sat enthroned behind the counter a splendid mature Arlesienne, whom my companion and I agreed that it was a rare privilege to contemplate. There is no rule of good manners or morals which makes it improper, at a cafe, to fix one's eyes upon the *dame de comptoir*, the lady is, in the nature of things, a part of your *consommation*. We were therefore free to admire without restriction the handsomest person I had ever seen give change for a five-franc piece. She was a large quiet woman, who would never see forty again; of an intensely feminine type, yet wonderfully rich and robust, and full of a certain physical nobleness. Though she was not really old, she was antique, and she was very grave, even a little sad. She had the dignity of a Roman empress, and she handled coppers as if they had been stamped with the head of Caesar. I have seen washerwomen in the Trastevere who were perhaps as handsome as she; but even the head-dress of the Roman contadina contributes less to the dignity of the person born to wear it than the sweet and stately Arlesian cap, which sits at once aloft and on the back of the head; which is accompanied with a wide black bow covering a considerable part of the crown; and which, finally, accommodates itself indescribably well to the manner in which the tresses of the front are pushed behind the cars.

This admirable dispenser of lumps of sugar has distracted me a little; for I am still not sufficiently historical. Before going to the cafe I had dined, and before dining I had found time to go and look at the arena. Then it was that I discovered that Arles has no general physiognomy, and, except the delightful little church of Saint Trophimus, no architecture, and that the rugosities of its dirty lanes affect the feet like knife-blades. It was not then, on the other hand, that I saw the arena best. The second day of my stay at Arles I devoted to a pilgrimage to the strange old hill town of Les Baux, the mediaeval Pompeii, of which I shall give myself the pleasure of speaking. The evening of that day, however (my friend and I returned in time for a late dinner), I wandered among the Roman remains of the place by the light of a magnificent moon, and gathered an impression which has lost little of its silvery glow. The moon of the evening before had been aqueous and erratic; but if on the present occasion it was guilty of any irregularity, the worst it did was only to linger beyond its time in the heavens, in order to let us look at things comfortably. The effect was admirable; it brought back the impression of the way, in Rome itself, on evenings like that, the moonshine rests upon broken shafts and slabs of antique pavement. As we sat in the theatre, looking at the two lone columns that survive part of the decoration of the back of the stage and at the fragments of ruin around them, we might have been in the Roman forum. The arena at Arles, with its great magnitude, is less complete than that of

Nimes; it has suffered even more the assaults of time and of the children of time, and it has been less repaired. The seats are almost wholly wanting; but the external walls minus the topmost tier of arches, are massively, ruggedly, complete; and the vaulted corridors seem as solid as the day they were built. The whole thing is superbly vast, and as monumental, for place of light amusement what is called in America a "varietyshow" as it entered only into the Roman mind to make such establishments. The *podium* is much higher than at Nimes, and many of the great white slabs that faced it have been recovered and put into their places. The proconsular box has been more or less reconstructed, and the great converging passages of approach to it are still majestically distinct: so that, as I sat there in the moon-charmed stillness, leaning my elbows on the battered parapet of the ring, it was not impossible to listen to the murmurs and shudders, the thick voice of the circus, that died away fifteen hundred years ago.

The theatre has a voice as well, but it lingers on the ear of time with a different music. The Roman theatre at Arles seemed to me one of the most charming and touching ruins I had ever beheld; I took a particular fancy to it. It is less than a skeleton, the arena may be called a skeleton; for it consists only of half a dozen bones. The traces of the row of columns which formed the scene the permanent back-scene remain; two marble pillars I just mentioned them are upright, with a fragment of their entablature. Be fore them is the vacant space which was filled by the stage, with the line of the prosoenium distinct, marked by a deep groove, impressed upon slabs of stone, which looks as if the bottom of a high screen had been intended to fit into it. The semicircle formed by the seats half a cup rises opposite; some of the rows are distinctly marked. The floor, from the bottom of the stage, in the shape of an arc of which the chord is formed by the line of the orchestra, is covered by slabs of colored marble red, yellow, and green which, though terribly battered and cracked to-day, give one an idea of the elegance of the interior. Everything shows that it was on a great scale: the large sweep of its enclosing walls, the massive corridors that passed behind the auditorium, and of which we can still perfectly take the measure. The way in which every seat commanded the stage is a lesson to the architects of our epoch, as also the immense size of the place is a proof of extraordinary power of voice on the part of the Roman actors. It was after we had spent half an hour in the moonshine at the arena that we came on to this more ghostly and more exquisite ruin. The principal entrance was locked, but we effected an easy *escalade*, scaled a low parapet, and descended into the place behind file scenes. It was as light as day, and the solitude was complete. The two slim columns, as we sat on the broken benches, stood there like a pair of silent actors. What I called touching, just now, was the thought that here the human voice, the utterance of a great language, had been supreme. The air was full of intonations and cadences; not of the echo of smashing blows, of riven armor, of howling victims and roaring beasts. The spot is, in short, one of the sweetest legacies of the ancient world; and there seems no profanation in the fact that by day it is open to the good people of Arles, who use it to pass, by no means in great numbers, from one part of the town to the other; treading the old marble floor, and brushing, if need be, the empty benches. This familiarity does not kill the place again; it makes it, on the contrary, live a little, makes the present and the past touch each other.

31

The third lion of Arles has nothing to do with the ancient world, but only with the old one. The church of Saint Trophimus, whose wonderful Romanesque porch is the principal ornament of the principal *place*, a *place* otherwise distinguished by the presence of a slim and tapering obelisk in the middle, as well as by that of the Hotel de Ville and the museum the interesting church of Saint Trophimus swears a little, as the French say, with the peculiar character of Arles. It is very remarkable, but I would rather it were in another place. Arles is delightfully pagan, and Saint Trophimus, with its apostolic sculptures, is rather a false note. These sculptures are equally remarkable for their primitive vigor and for the perfect preservation in which they have come down to us. The deep recess of a round-arched porch of the twelfth century is covered with quaint figures, which have not lost a nose or a finger. An angular, Byzantine-looking Christ sits in a diamond-shaped frame at the summit of the arch, surrounded by little angels, by great apostles, by winged beasts, by a hundred sacred symbols and grotesque ornaments. It is a dense embroidery of sculpture, black with time, but as uninjured as if it had been kept under glass. One good mark for the French Revolution! Of the interior of the church, which has a nave of the twelfth century, and a choir three hundred years more recent, I chiefly remember the odd feature that the Romanesque aisles are so narrow that you literally or almost squeeze through them. You do so with some eagerness, for your natural purpose is to pass out to the cloister. This cloister, as distinguished and as perfect as the porch, has a great deal of charm. Its four sides, which are not of the same period (the earliest and best are of the twelfth century), have an elaborate arcade, supported on delicate pairs of columns, the capitals of which show an extraordinary variety of device and ornament. At the corners of the quadrangle these columns take the form of curious human figures. The whole thing is a gem of lightness and preservation, and is often cited for its beauty; but if it doesn't sound too profane I prefer, especially at Arles, the ruins of the Roman theatre. The antique element is too precious to be mingled with anything less rare. This truth was very present to my mind during a ramble of a couple of hours that I took just before leaving the place; and the glowing beauty of the morning gave the last touch of the impression. I spent half an hour at the Museum; then I took another look at the Roman theatre; after which I walked a little out of the town to the Aliscamps, the old Elysian Fields, the meager remnant of the old pagan place of sepulture, which was afterwards used by the Christians, but has been for ages deserted, and now consists only of a melancholy avenue of cypresses, lined with a succession of ancient sarcophagi, empty, mossy, and mutilated. An iron-foundry, or some horrible establishment which is conditioned upon tall chimneys and a noise of hammering and banging, has been established near at hand; but the cypresses shut it out well enough, and this small patch of Elysium is a very romantic corner.

The door of the Museum stands ajar, and a vigilant custodian, with the usual batch of photographs on his mind, peeps out at you disapprovingly while you linger opposite, before the charming portal of Saint Trophimus, which you may look at for nothing. When you succumb to the silent influence of his eye, and go over to visit his collection, you find yourself in a desecrated church, in which a variety of ancient objects, disinterred in Arlesian soil, have been arranged without any pomp. The best of these, I believe, were found in the ruins of the theatre. Some of the most curious of them are early Christian sarcophagi, exactly on the pagan model, but covered with rude yet vigorously wrought images of the apostles, and with illustrations of scriptural history. Beauty of the highest kind, either of conception or of execution, is absent from most of the Roman fragments, which belong to the taste of a late period and a provincial civilization. But a gulf divides them from the bristling little imagery of the Christian sarcophagi, in which, at the same time, one detects a vague emulation of the rich examples by which their authors were surrounded. There is a certain element of style in all the pagan things; there is not a hint of it in the early Christian relics, among which, according to M. Joanne, of the Guide, are to be found more fine sarcophagi than in any collection but that of St. John Lateran. In two or three of the Roman fragments there is a noticeable distinction; principally in a charming bust of a boy, quite perfect, with those salient eyes that one sees in certain antique busts, and to which the absence of vision in the marble mask gives a look, often very touching, as of a baffled effort to see; also in the head of a woman, found in the ruins of the theatre, who, alas! has lost her nose, and whose noble, simple contour, barring this deficiency, recalls the great manner of the Venus of Milo. There are various rich architectural fragments which indicate that that edifice was a very splendid affair. This little Museum at Arles, in short, is the most Roman thing I know of, out of Rome.

32

I find that I declared one evening, in a little journal I was keeping at that time, that I was weary of writing (I was probably very sleepy), but that it was essential I should make some note of my visit to Les Baux. I must have gone to sleep as soon as I had recorded this necessity, for I search my small diary in vain for any account of that enchanting spot. I have nothing but my memory to consult, a memory which is fairly good in regard to a general impression, but is terribly infirm in the matter of details and items. We knew in advance, my companion and I that Les Baus was a pearl of picturesqueness; for had we not read as much in the handbook of Murray, who has the testimony of an English nobleman as to its attractions? We also knew that it lay some miles from Aries, on the crest of the Alpilles, the craggy little mountains which, as I stood on the breezy platform of Beaucaire, formed to my eye a charming, if somewhat remote, background to Tarascon; this assurance having been given us by the landlady of the inn at Arles, of whom we hired a rather lumbering conveyance. The weather was not promising, but it proved a good day for the mediaeval Pompeii; a gray, melancholy, moist, but rainless, or almost rainless day, with nothing in the sky to flout, as the poet says, the dejected and pulverized past. The drive itself was charming; for there is an inexhaustible sweetness in the gray-green landscape of Provence. It is never absolutely flat, and yet is never really ambitious, and is full both of entertainment and repose. It is in constant undulation, and the bareness of the soil lends itself easily to outline and profile. When I say the bareness, I mean the absence of woods and hedges. It blooms with heath and scented shrubs and stunted olive; and the white rock shining through the scattered herbage has a brightness which answers to the brightness of the sky. Of course it needs the sunshine, for all southern countries look a little false under the ground glass of incipient bad weather. This was the case on the day of my pilgrimage to Les Baux. Nevertheless, I was as glad to keep going as I was to arrive; and as I went it seemed to me that true happiness would consist in wandering through such a land on foot, on September afternoons, when one might stretch one's self on the warm ground in some shady hollow, and listen to the hum of bees and the whistle of melancholy shepherds; for in Provence the shepherds whistle to their flocks. I saw two or three of them, in the course of this drive to Les Baux, meandering about, looking behind, and calling upon the sheep in this way to follow, which the sheep always did, very promptly, with ovine unanimity. Nothing is more picturesque than to see a slow shepherd threading his way down one of the winding paths on a hillside, with his flock close behind him, necessarily expanded, yet keeping just at his heels, bending and twisting as it goes, and looking rather like the tail of a dingy comet.

About four miles from Arles, as you drive northward toward the Alpilles, of which Alphonse Daudet has spoken so often, and, as he might say, so intimately, stand on a hill that overlooks the road the very considerable ruins of the abbey of Montmajour, one of the innumerable remnants of a feudal and ecclesiastical (as well as an architectural) past that one encounters in the South of France; remnants which, it must be confessed, tend to introduce a certain confusion and satiety into the passive mind of the tourist. Montmajour, however, is very impressive and interesting; the only trouble with it is that, unless you have stopped and returned to Arles, you see it in memory over the head of Les Baux, which is a much more absorbing picture. A part of the mass of buildings (the monastery) dates only from the last century; and the stiff architecture of that period does not lend itself very gracefully to desolation: it looks too much as if it had been burnt down the year before. The monastery was demolished during the Revolution, and it injures a little the effect of the very much more ancient fragments that are connected with it. The whole place is on a great scale; it was a rich and splendid abbey. The church, a vast basilica of the eleventh century, and of the noblest proportions, is virtually intact; I mean as regards its essentials, for the details have completely vanished. The huge solid shell is full of expression; it looks as if it had been hollowed out by the sincerity of early faith, and it opens into a cloister as impressive as itself. Wherever one goes, in France, one meets, looking backward a little, the specter of the great Revolution; and one meets it always in the shape of the destruction of something beautiful and precious. To make us forgive it at all, how much it must also have destroyed that was more hateful than itself! Beneath the church of Montmajour is a most extraordinary crypt, almost as big as the edifice above it, and making a complete subterranean temple, surrounded with a circular gallery, or deambulatory, which expands it intervals into five square chapels. There are other things, of which I have but a confused memory: a great fortified keep; a queer little primitive chapel, hollowed out of the rock, beneath these later structures, and recommended to the visitor's attention as the confessional of Saint Trophimus, who shares with so many worthies the glory of being the first apostle of the Gauls. Then there is a strange, small church, of the dimmest antiquity, standing at a distance from the other buildings. I remember that after we had let ourselves down a good many steepish places to visit crypts and confessionals, we walked across a field to this archaic cruciform edifice, and went thence to a point further down the road, where our carriage was awaiting us. The chapel of the Holy Cross, as it is called, is classed among the historic monuments of France; and I read in a queer, rambling, ill-written book which I picked up at Avignon, and in which the author, M. Louis de Lainbel, has buried a great deal of curious information on the subject of Provence, under a style inspiring little confidence, that the "delicieuse chapelle de Sainte-Croix" is a "veritable bijou artistique." He speaks of "a piece of lace in stone," which runs from one end of the building to the other, but of which I am obliged to confess that I have no recollection. I retain, however, a sufficiently clear impression of the little superannuated temple, with its four apses and its perceptible odor of antiquity, the odor of the eleventh century.

The ruins of Les Baux remain quite indistinguishable, even when you are directly beneath them, at the foot of the charming little Alpilles, which mass themselves with a kind of delicate ruggedness. Rock and ruin have been so welded together by the confusions of time, that as you approach it from behind that is, from the direction of Arles the place presents simply a general air of cragginess. Nothing can be prettier than the crags of Provence; they are beautifully modeled, as painters say, and they have a delightful silvery color. The road winds round the foot of the hills on the top of which Lea Baux is planted, and passes into another valley, from which the approach to the town is many degrees less precipitous, and may be comfortably made in a carriage. Of course the deeply inquiring traveler will alight as promptly as possible; for the pleasure of climbing into this queerest of cities on foot is not the least part of the entertainment of going there. Then you appreciate its extraordinary position, its picturesqueness, its steepness, its desolation and decay. It hangs that is, what remains of it to the slanting summit of the mountain. Nothing would be more natural than for the whole place to roll down into the valley. A part of it has done so for it is not unjust to suppose that in the process of decay the crumbled particles have sought the lower level; while the remainder still clings to its magnificent perch.

If I called Les Baux a city, just, above, it was not that I was stretching a point in favor of the small spot which to-day contains but a few dozen inhabitants. The history of the plate is as extraordinary as its situation. It was not only a city, but a state; not only a state, but an empire; and on the crest of its little mountain called itself sovereign of a territory, or at least of scattered towns and counties, with which its present aspect is grotesquely out of relation. The lords of Les Baux, in a word, were great feudal proprietors; and there-was a time during which the island of Sardinia, to say nothing of places nearer home, such as Arles and Marseilles, paid them homage. The chronicle of this old Provencal house has been written, in a style somewhat unctuous and flowery, by M. Jules Canonge. I purchased the little book a modest pamphlet at the establishment of the good sisters, just beside the church, in one of the highest parts of Les Baux. The sisters have a school for the hardy little Baussenques, whom I heard piping their lessons, while I waited in the cold *parloir* for one of the ladies to come and speak to me. Nothing could have been more perfect than the manner of this excellent woman when she arrived; yet her small religious house seemed a very out-of-the-way corner of the world. It was spotlessly neat, and the rooms looked as if they had lately been papered and painted: in this respect, at the mediaeval Pompeii, they were rather a discord. They were, at any rate, the newest, freshest thing at Les Baux. I remember going round to the church, after I had left the good sisters, and to a little quiet terrace, which stands in front of it, ornamented with a few small trees and bordered with a wall, breasthigh, over which you look down steep hillsides, off into the air and all about the neighboring country. I remember saying to myself that this little terrace was one of those felicitous nooks which the tourist of taste keeps in his mind as a picture. The church was small and brown and dark, with a certain rustic richness. All this, however, is no general description of Les Baux.

I am unable to give any coherent account of the place, for the simple reason that it is a mere confusion of ruin. It has not been preserved in lava like Pompeii, and its streets and houses, its ramparts and castle, have become fragmentary, not through the sudden destruction, but through the gradual withdrawal, of a population. It is not an extinguished, but a deserted city; more deserted far than even Carcassonne and Aigues-Mortes, where I found so much entertainment in the grass-grown element. It is of very small extent, and even in the days of its greatness, when its lords entitled themselves counts of Cephalonia and Neophantis, kings of Arles and Vienne, princes of Achaia, and emperors of Constantinople, even at this flourishing period, when, as M. Jules Canonge remarks, "they were able to depress the balance in which the fate of peoples and kings is weighed," the plucky little city contained at the most no more than thirty-six hundred souls. Yet its lords (who, however, as I have said, were able to present a long list of subject towns, most of them, though a few are renowned, unknown to fame) were seneschals and captains-general of Piedmont and Lombardy, grand admirals of the kingdom of Naples, and its ladies were sought in marriage by half the first princes in Europe. A considerable part of the little narrative of M. Canonge is taken up with the great alliances of the House of Baux, whose fortunes, matrimonial and other, he traces from the eleventh century down to the sixteenth. The empty shells of a considerable number of old houses, many of which must have been superb, the lines of certain steep little streets, the foundations of a castle, and ever so many splendid views, are all that remains to-day of these great titles. To such a list I may add a dozen very polite and sympathetic people, who emerged from the interstices of the desultory little town to gaze at the two foreigners who had driven over from Arles, and whose horses were being baited at the modest inn. The resources of this establishment we did not venture otherwise to test, in spite of the seductive fact that the sign over the door was in the Provencal tongue. This little group included the baker, a rather melancholy young man, in high boots and a cloak, with whom and his companions we had a good deal of conversation. The Baussenques of to-day struck me as a very mild and agreeable race, with a good deal of the natural amenity which, on occasions like this one, the traveler, who is, waiting for his horses to be put in or his dinner to be prepared, observes in the charming people who lend themselves to conversation in the hill-towns of Tuscany. The spot where our entertainers at Les Baux congregated was naturally the most inhabited portion of the town; as I say, there were at least a dozen human figures within sight. Presently we wandered away from them, scaled the higher places, seated ourselves among the ruins of the castle, and looked down from the cliff overhanging that portion of the road which I have mentioned as approaching Les Baux from behind. I was unable to trace the configuration of the castle as plainly as the writers who have described it in the guide-books, and I am ashamed to say that I did not even perceive the three great figures of stone (the three Marys, as they are called; the two Marys of Scripture, with Martha), which constitute one of the curiosities of the place, and of which M. Jules Canonge speaks with almost hyperbolical admiration. A brisk shower, lasting some ten minutes, led us to take refuge in a cavity, of mysterious origin, where the melancholy baker presently discovered us, having had the *bonne pensee* of coming up for us with an umbrella which certainly belonged, in former ages, to one of the Stephanettes or Berangeres commemorated by M. Canonge. His oven, I am afraid, was cold so long as our visit lasted. When the rain was over we wandered down to the little disencumbered space before the inn, through a small labyrinth of obliterated things. They took the form of narrow, precipitous streets, bordered by empty houses, with gaping windows and absent doors, through which we had glimpses of sculptured chimney-pieces and fragments of stately arch and vault. Some of the houses are still inhabited; but most of them are open to the air and weather. Some of them have completely collapsed; others present to the street a front which enables one to judge of the physiognomy of Les Baux in the days of its importance. This importance had pretty well passed away in the early part of the sixteenth century, when the place ceased to be an independent principality. It became by bequest of one of its lords, Bernardin des Baux, a great captain of his time part of the appanage of the kings of France, by whom it was placed under the protection of Arles, which had formerly occupied with regard to it a different position. I know not whether the Arlesians neglected their trust; but the extinction of the sturdy little stronghold is too complete not to have begun long ago. Its memories are buried under its ponderous stones. As we drove away from it in the gloaming, my friend and I agreed that the two or three hours we had spent there were among the happiest impressions of a pair of tourists very curious in the picturesque. We almost forgot that we were bound to regret that the shortened day left us no time to drive five miles further, above a pass in the little mountains it had beckoned to us in the morning, when we came in sight of it, almost irresistibly to see the Roman arch and mausoleum of Saint Remy. To compass this larger excursion (including the visit to Les Baux) you must start from Arles very early in the morning; but I can imagine no more delightful day.

33

I had been twice at Avignon before, and yet I was not satisfied. I probably am satisfied now; nevertheless, I enjoyed my third visit. I shall not soon forget the first, on which a particular emotion set indelible stamp. I was travelling northward, in 1870, after four months spent, for the first time, in Italy. It was the middle of January, and I had found myself, unexpectedly, forced to return to England for the rest of the winter. It was an insufferable disappointment; I was wretched and broken-hearted. Italy appeared to me at that time so much better than anything else in the world, that to rise from table in the middle of the feast was a prospect of being hungry for the rest of my days. I had heard a great deal of praise of the south of France; but the south of France was a poor consolation. In this state of mind I arrived at Avignon, which under a bright, hard winter sun was tingling fairly spinning with the *mistral*. I find in my journal of the other day a reference to the acuteness of my reluctance in January, 1870. France, after Italy, appeared, in the language of the latter country, *poco simpatica*, and I thought it necessary, for reasons now inconceivable, to read the "Figaro," which was filled with descriptions of the horrible Troppmann, the murderer of the *famille* Kink. Troppmann, Kink, *le crime do Pantin*, very names that figured in this episode seemed to wave me back. Had I abandoned the sonorous south to associate with vocables so base?

It was very cold, the other day, at Avignon; for though there was no mistral, it was raining as it rains in Provence, and the dampness had a terrible chill in it. As I sat by my fire, late at night for in genial Avignon, in October, I had to have a fire it came back to me that eleven years before I had at that same hour sat by a fire in that same room, and, writing to a friend to whom I was not afraid to appear extravagant, had made a vow that at some happier period of the future I would avenge myself on the *cidevant* city of the Popes by taking it in a contrary sense. I suppose that I redeemed my vow on the occasion of my second visit better than on my third; for then I was on my way to Italy, and that vengeance, of course, was complete. The only drawback was that I was in such a hurry to get to Ventimiglia (where the Italian custom-house was to be the sign of my triumph), that I scarcely took time to make it clear to myself at Avignon that this was better than reading the "Figaro." I hurried on almost too fast to enjoy the consciousness of moving southward. On this last occasion I was unfortunately destitute of that happy faith. Avignon was my southernmost limit; after which I was to turn round and proceed back to England. But in the interval I had been a great deal in Italy, and that made all the difference.

I had plenty of time to think of this, for the rain kept me practically housed for the first twenty-four hours. It had been raining in, these regions for a month, and people had begun to look askance at the Rhone, though as yet the volume of the river was not exorbitant. The only excursion possible, while the torrent descended, was a kind of horizontal dive, accompanied with infinite splashing, to the little *musee* of the town, which is within a moderate walk of the hotel. I had a memory of it from my first visit; it had appeared to me more pictorial than its pictures. I found that recollection had flattered it a little, and that it is neither better nor worse than most provincial museums. It has the usual musty chill in the air, the usual grass-grown fore-court, in which a few lumpish Roman fragments are disposed, the usual red tiles on the floor, and the usual specimens of the more livid schools on the walls. I rang up the *gardien*, who arrived with a bunch of keys, wiping his mouth; he unlocked doors for me, opened shutters, and while (to my distress, as if the things had been worth lingering over) he shuffled about after me, he announced the names of the pictures before which I stopped, in a voice that reverberated through the melancholy halls, and seemed to make the authorship shameful when it was obscure, and grotesque when it pretended to be great. Then there were intervals of silence, while I stared absent-mindedly, at hap-hazard, at some

132

indistinguishable canvas, and the only sound was the downpour of the rain on the skylights. The museum of Avignon derives a certain dignity from its Roman fragments. The town has no Roman monuments to show; in this respect, beside its brilliant neighbors, Arles and Nimes, it is a blank. But a great many small objects have been found in its soil, pottery, glass, bronzes, lamps, vessels and ornaments of gold and silver. The glass is especially chaming, small vessels of the most delicate shape and substance, many of them perfectly preserved. These diminutive, intimate things bring one near to the old Roman life; they seem like pearls strung upon the slender thread that swings across the gulf of time. A little glass cup that Roman lips have touched says more to us than the great vessel of an arena. There are two small silver *casseroles*, with chiseled handles, in the museum of Avignon, that struck me as among the most charming survivals of antiquity.

I did wrong just above, to speak of my attack on this establishment as the only recreation I took that first wet day; for I remember a terribly moist visit to the former palace of the Popes, which could have taken place only in the same tempestuous hours. It is true that I scarcely know why I should have gone out to see the Papal palace in the rain, for I had been over it twice before, and even then had not found the interest of the place so complete as it ought to be; the fact, nevertheless, remains that this last occasion is much associated with an umbrella, which was not superfluous even in some of the chambers and corridors of the gigantic pile. It had already seemed to me the dreariest of all historical buildings, and my final visit confirmed the impression. The place is as intricate as it is vast, and as desolate as it is dirty. The imagination has, for some reason or other, to make more than the effort usual in such cases to restore and repeople it. The fact, indeed, is simply that the palace has been so incalculably abused and altered. The alterations have been so numerous that, though I have duly conned the enumerations, supplied in guidebooks, of the principal perversions, I do not pretend to carry any of them in my head. The huge bare mass, without ornament, without grace, despoiled of its battlements and defaced with sordid modern windows, covering the Rocher des Doms, and looking down over the Rhone and the broken bridge of Saint-Benazet (which stops in such a sketchable manner in midstream), and across at the lonely tower of Philippe le Bel and the ruined wall of Villeneuve, makes at a distance, in spite of its poverty, a great figure, the effect of which is carried out by the tower of the church beside it (crowned though the latter be, in a top-heavy fashion, with an immense modern image of the Virgin) and by the thick, dark foliage of the garden laid out on a still higher portion of the eminence. This garden recalls, faintly and a trifle perversely, the grounds of the Pincian at Rome. I know not whether it is the shadow of the Papal name, present in both places, combined with a vague analogy between the churches, which, approached in each case by a flight of steps, seemed to defend the precinct, but each time I have seen the Promenade des Doms it has carried my thoughts to the wider and loftier terrace from which you look away at the Tiber and Saint Peter's.

As you stand before the Papal palace, and especially as you enter it, you are struck with its being a very dull monument. History enough was enacted here: the great schism lasted from 1305 to 1370, during which seven Popes, all Frenchmen, carried on the court of Avignon on principles that have not commended themselves to the esteem of posterity. But history has been whitewashed away, and the scandals of that period have mingled with the dust of dilapidations and repairs. The building has for many years been occupied as a barrack for regiments of the line, and the main characteristics of a barrack an extreme nudity and a very queer smell prevail throughout its endless compartments. Nothing could have been more cruelly dismal than the appearance it presented at the time of this third visit of mine. A regiment, changing quarters, had departed the day before, and another was expected to arrive (from Algeria) on the morrow. The place had been left in the befouled and belittered condition which marks the passage of the military after they have broken carnp, and it would offer but a melancholy welcome to the regiment that was about to take possession. Enormous windows had been left carelessly open all over the building, and the rain and wind were beating into empty rooms and passages; making draughts which purified, perhaps, but which scarcely cheered. For an arrival, it was horrible. A handful of soldiers had remained behind. In one of the big vaulted rooms several of them were lying on their wretched beds, in the dim light, in the cold, in the damp, with the bleak, bare walls before them, and their overcoats, spread over them, pulled up to their noses. I pitied them immensely, though they may have felt less wretched than they looked. I thought not of the old profligacies and crimes, not of the funnel-shaped torture-chamber (which, after exciting the shudder of generations, has been ascertained now, I believe, to have been a mediaeval bakehouse), not of the tower of the *glaciere* and the horrors perpetrated here in the Revolution, but of the military burden of young France. One wonders how young France endures it, and one is forced to believe that the French conscript has, in addition to his notorious good-humor, greater toughness than is commonly supposed by those who consider only the more relaxing influences of French civilization. I hope he finds occasional compensation for such moments as I saw those damp young peasants passing on the mattresses of their hideous barrack, without anything around to remind them that they were in the most civilized of countries. The only traces of former splendor now visible in the Papal pile are the walls and vaults of two small chapels, painted in fresco, so battered and effaced as to be scarcely distinguishable, by Simone Memmi. It offers, of course, a peculiarly good field for restoration, and I believe the government intend to take it in hand. I mention this fact without a sigh; for they cannot well make it less interesting than it is at present.

34

Fortunately, it did not rain every day (though I believe it was raining everywhere else in the department); otherwise I should not have been able to go to Villeneuve and to Vaucluse. The afternoon, indeed, was lovely when I walked over the interminable bridge that spans the two arms of the Rhone, divided here by a considerable island, and directed my course, like a solitary horseman on foot, to the lonely tower which forms one of the outworks of Villeneuve-les-Avignon. The picturesque, half-deserted little town lies a couple of miles further up the river. The immense round towers of its old citadel and the long stretches of ruined wall covering the slope on which it lies, are the most striking features of the nearer view, as you look from Avignon across the Rhone. I spent a couple of hours in visiting these objects, and there was a kind of pictorial sweetness in the episode; but I have not many details to relate. The isolated tower I just mentioned has much in common with the detached donjon of Montmajour, which I had looked at in going to Les Baux, and to which I paid my respects in speaking of that excursion. Also the work of Philippe le Bel (built in 1307), it is amazingly big and stubborn, and formed the opposite limit of the broken bridge, whose first arches (on the side of Avignon) alone remain to give a measure of the occasional volume of the Rhone. Half an hour's walk brought me to Villeneuve, which lies away from the river, looking like a big village, half depopulated, and occupied for the most part by dogs and cats, old women and small children; these last, in general, remarkably pretty, in the manner of the children of Provence. You pass through the place, which seems in a singular degree vague and unconscious, and come to the rounded hill on which the ruined abbey lifts its yellow walls, the Benedictine abbey of Saint Andre, at once a church, a monastery, and a fortress. A large part of the crumbling enceinte disposes itself over the hill; but for the rest, all that has preserved any traceable cohesion is a considerable portion, of the citadel. The defense of the place appears to have been entrusted largely to the huge round towers that flank the old gate; one of which, the more complete, the ancient warden (having first inducted me into his own dusky little apartment, and presented me with a great bunch of lavender) enabled me to examine in detail. I would almost have dispensed with the privilege, for I think I have already mentioned that an acquaintance with many feudal interiors has wrought a sad confusion in my mind. The image of the outside always remains distinct; I keep it apart from other images of the same sort; it makes a picture sufficiently ineffaceable. But the guard-rooms, winding staircases, loop-holes, prisons, repeat themselves and intermingle; they have a wearisome family likeness. There are always black passages and corners, and walls twenty feet thick; and there is always some high place to climb up to for the sake of a "magnificent" view. The views, too, are apt to get muddled. These dense gate-towers of Philippe le Bel struck me, however, as peculiarly wicked and grim. Their capacity is of the largest, and they contain over so many devilish little dungeons, lighted by the narrowest slit in the prodigious wall, where it comes over one with a good deal of vividness and still more horror that wretched human beings ever lay there rotting in the dark. The dungeons of Villeneuve made a particular impression on me, greater than any, except those of Loches, which must surely be the most grewsome in Europe. I hasten to add that every dark hole at Villeneuve is called a dungeon; and I believe it is well established that in this manner, in almost all old castles and towers, the sensibilities of the modern tourist are unscrupulously played upon. There were plenty of black holes in the Middle Ages that were not dungeons, but household receptacles of various kinds; and many a tear dropped in pity for the groaning captive has really been addressed to the spirits of the larder and the faggot-nook. For all this, there are some very bad corners in the towers of Villeneuve, so that I was not wide of the mark when I began to think again, as I had often thought before, of the stoutness

136

of the human composition in the Middle Ages, and the tranquility of nerve of people to whom the groaning captive and the blackness of a "living tomb" were familiar ideas, which did not at all interfere with their happiness or their sanity. Our modern nerves, our irritable sympathies, our easy discomforts and fears, make one think (in some relations) less respectfully of human nature. Unless, indeed, it be true, as I have heard it maintained, that in the Middle Ages every one did go mad, every one *was* mad. The theory that this was a period of general insanity is not altogether indefensible.

Within the old walls of its immense abbey the town of Villeneuve has built itself a rough faubourg; the fragments with which the soil was covered having been, I suppose, a quarry of material. There are no streets; the small, shabby houses, almost hovels, straggle at random over the uneven ground. The only important feature is a convent of cloistered nuns, who have a large garden (always within the walls) behind their house, and whose doleful establishment you look down into, or down at simply, from the battlements of the citadel. One or two of the nuns were passing in and out of the house; they wore gray robes, with a bright red cape. I thought their situation most provincial. I came away, and wandered a little over the base of the hill, outside the walls. Small white stones cropped through the grass, over which low olive-trees were scattered. The afternoon had a yellow brightness. I sat down under one of the little trees, on the grass, the delicate gray branches were not much above my head, and rested, and looked at Avignon across the Rhone. It was very soft, very still and pleasant, though I am not sure it was all I once should have expected of that combination of elements: an old city wall for a background, a canopy of olives, and, for a couch, the soil of Provence.

When I came back to Avignon the twilight was already thick; but I walked up to the Rocher des Doms. Here I again had the benefit of that amiable moon which had already lighted up for me so many romantic scenes. She was full, and she rose over the Rhone, and made it look in the distance like a silver serpent. I remember saying to myself at this moment, that it would be a beautiful evening to walk round the walls of Avignon, the remarkable walls, which challenge comparison with those of Carcassonne and Aigues-Mortes, and which it was my duty, as an observer of the picturesque, to examine with some attention. Presenting themselves to that silver sheen, they could not fail to be impressive. So, at least, I said to myself; but, unfortunately, I did not believe what I said. It is a melancholy fact that the walls of Avignon had never impressed me at all, and I had never taken the trouble to make the circuit. They are continuous and complete, but for some mysterious reason they fail of their effect. This is partly because they are very low, in some places almost absurdly so; being buried in new accumulations of soil, and by the filling in of the moat up to their middle. Then they have been too well tended; they not only look at present very new, but look as if they had never been old. The fact that their extent is very much greater makes them more of a curiosity than those of Carcassonne; but this is exactly, as the same time, what is fatal to their pictorial unity. With their thirty-seven towers and seven gates they lose themselves too much to make a picture that will compare with the admirable little vignette of Carcassonne. I may mention, now that I am speaking of the general mass of Avignon, that nothing is more curious than the way in which, viewed from a distance, it is all reduced to naught by the vast bulk of the palace of the Popes. From across the Rhone, or from the train, as you leave the place, this great gray block is all Avignon; it seems to occupy the whole city, extensive, with its shrunken population, as the city is.

35

It was the morning after this, I think (a certain Saturday), that when I came out of the Hotel de l'Europe, which lies in a shallow concavity just within the city gate that opens on the Rhone, came out to look at the sky from the little *place* before the inn, and see how the weather promised for the obligatory excursion to Vaucluse, I found the whole town in a terrible taking. I say the whole town advisedly; for every inhabitant appeared to have taken up a position on the bank of the river, or on the uppermost parts of the promenade of the Doms, where a view of its course was to be obtained. It had risen surprisingly in the night, and the good people of Avignon had reason to know what a rise of the Rhone might signify. The town, in its lower portions, is quite at the mercy of the swollen waters; and it was mentioned to me that in 1856 the Hotel de l'Europe, in its convenient hollow, was flooded up to within a few feet of the ceiling of the dining-room, where the long board which had served for so many a table d'hote floated disreputably, with its legs in the air. On the present occasion the mountains of the Ardeche, where it had been raining for a month, had sent down torrents which, all that fine Friday night, by the light of the innocent-looking moon, poured themselves into the Rhone and its tributary, the Durance. The river was enormous, and continued to rise; and the sight was beautiful and horrible. The water in many places was already at the base of the city walls; the quay, with its parapet just emerging, being already covered. The country, seen from the Plateau des Doms, resembled a vast lake, with protrusions of trees, houses, bridges, gates. The people looked at it in silence, as I had seen people before on the occasion of a rise of the Arno, at Pisa appear to consider the prospects of an inundation. "Il monte; il monte toujours," there was not much said but that. It was a general holiday, and there was an air of wishing to profit, for sociability's sake, by any interruption of the commonplace (the popular mind likes "a change," and the element of change mitigates the sense of disaster); but the affair was not otherwise a holiday. Suspense and anxiety were in the air, and it never is pleasant to be reminded of the helplessness of man. In the presence of a loosened river, with its ravaging, unconquerable volume, this impression is as strong as possible; and as I looked at the deluge which threatened to make an island of the Papal palace, I perceived that the scourge of water is greater than the scourge of fire. A blaze may be quenched, but where could the flame be kindled that would arrest the quadrupled Rhone? For the population of Avignon a good deal was at stake, and I am almost ashamed to confess that in the midst of the public alarm I considered the situation from the point of view of the little projects of a sentimental tourist. Would the prospective inundation interfere with my visit to Vaucluse, or make it imprudent to linger twenty-four hours longer at Avignon? I must add that the tourist was not perhaps, after all, so sentimental. I have spoken of the pilgrimage to the shrine of Petrarch as obligatory, and that was, in fact, the light in which it presented itself to me; all the more that I had been twice at Avignon without undertaking it. This why I was vexed at the Rhone if vexed I was for representing as impracticable an excursion which I cared nothing about. How little I cared was manifest from my inaction on former occasions. I had a prejudice against Vancluse, against Petrarch, even against the incomparable Laura. I was sure that the place was cockneyfied and threadbare, and I had never been able to take an interest in the poet and the lady. I was sure that I had known many women as charming and as handsome as she, about whom much less noise had been made; and I was convinced that her singer was factitious and literary, and that there are half a dozen stanzas in Wordsworth that speak more to the soul than the whole collection of his *fioriture*. This was the crude state of mind in which I determined to go, at any risk, to Vaucluse. Now that I think it over, I seem to remember that I had hoped, after all, that the submersion of the roads would forbid it. Since morning the clouds had gathered again, and by noon they were so heavy that there was every prospect of a torrent. It appeared absurd to choose such a time as this to visit a fountain a fountain which, would be indistinguishable in the general cataract. Nevertheless I took a vow that if at noon the rain should not have begun to descend upon Avignon I would repair to the head-spring of the Sorgues. When the critical moment arrived, the clouds were hanging over Avignon like distended water-bags, which only needed a prick to empty themselves. The prick was not given, however; all nature was too much occupied in following the aberration of the Rhone to think of playing tricks elsewhere. Accordingly, I started for the station in a spirit which, for a tourist who sometimes had prided himself on his unfailing supply of sentiment, was shockingly perfunctory.

"For tasks in hours of insight willed May be in hours of gloom fulfilled."

I remembered these lines of Matthew Arnold (written, apparently, in an hour of gloom), and carried out the idea, as I went, by hoping that with the return of insight I should be glad to have seen Vaucluse. Light has descended upon me since then, and I declare that the excursion is in every way to be recommended. The place makes a great impression, quite apart from Petrarch and Laura.

There was no rain; there was only, all the afternoon, a mild, moist wind, and a sky magnificently black, which made a *repoussoir* for the paler cliffs of the fountain. The road, by train, crosses a flat, expressionless country, toward the range of arid hills which lie to the east of Avignon, and which spring (says Murray) from the mass of the Mont-Ventoux. At Isle-sur-Sorgues, at the end of about an hour, the foreground becomes much more animated and the distance much more (or perhaps I should say much less) actual. I descended from the train, and ascended to the top of an omnibus which was to convey me into the recesses of the hills. It had not been among my previsions that I should be indebted to a vehicle of that kind for an opportunity to commune with the spirit of Petrarch; and I had to borrow what consolation I could from the fact that at least I had the omnibus to myself. I was the only passenger; every one else was at Avignon, watching the Rhone. I lost no time in perceiving that I could not have come to Vaucluse at a better moment. The Sorgues was almost as full as the Rhone, and of a color much more romantic. Rushing along its narrowed channel under an avenue of fine *platanes* (it is confined between solid little embankments of stone), with the good-wives of the village, on the brink, washing their linen in its contemptuous flood, it gave promise of high entertainment further on.

The drive to Vaucluse is of about three quarters of an hour; and though the river, as I say, was promising, the big pale hills, as the road winds into them, did not look as if their slopes of stone and shrub were a nestling-place for superior scenery. It is a part of the merit of Vaucluse, indeed, that it is as much as possible a surprise. The place has a right to its name, for the valley appears impenetrable until you get fairly into it. One perverse twist follows another, until the omnibus suddenly deposits you in front of the "cabinet" of Petrarch. After that you have only to walk along the left bank of the river. The cabinet of Petrarch is to-day a hideous little *cafe*, bedizened, like a signboard, with extracts from the ingenious "Rime." The poet and his lady are, of course, the stock in trade of the little village, which has had for several generations the privilege of attracting young couples engaged in their wedding-tour, and other votaries of the tender passion. The place has long been familiar, on festal Sundays, to the swains of Avignon and their attendant nymphs. The little fish of the Sorgues are much esteemed, and, eaten on the spot, they constitute, for the children of the once Papal city, the classic suburban dinner. Vaucluse has been turned to account, however, not only by sentiment, but by industry; the banks of the stream being disfigured by a pair of hideous mills for the manufacture of paper and of wool. In an enterprising and economical age the water-power of the Sorgues was too obvious a motive; and I must say that, as the torrent rushed past them, the wheels of the dirty little factories appeared to turn merrily enough. The footpath on the left bank, of which I just spoke, carries one, fortunately, quite out of sight of them, and out of sound as well, inasmuch as on the day of my visit the stream itself, which was in tremendous force, tended more and more, as one approached the fountain, to fill the valley with its own echoes. Its color was magnificent, and the whole spectacle more like a corner of Switzerland than a nook in Provence. The protrusions of the mountain shut it in, and you penetrate to the bottom of the recess which they form. The Sorgues rushes and rushes; it is almost like Niagara after the jump of the cataract. There are dreadful little booths beside the path, for the sale of photographs and *immortelles*, I don't know what one is to do with the immortelles, where you are offered a brush dipped in tar to write your name withal on the rocks. Thousands of vulgar persons, of both sexes, and exclusively, it appeared, of the French nationality, had availed themselves of this implement; for every square inch of accessible stone was scored over with some human appellation. It is not only we in America, therefore, who besmirch our scenery; the practice exists, in a more organized form (like everything else in France), in the country of good taste. You leave the little booths and stalls behind; but the bescribbled crag, bristling with human vanity, keeps you company even when you stand face to face with the fountain. This happens when you find yourself at the foot of the enormous straight cliff out of which the river gushes. It rears itself to an extraordinary height, a huge forehead of bare stone, looking as if it were the half of a tremendous mound, split open by volcanic action. The little valley, seeing it there, at a bend, stops suddenly, and receives in its arms the magical spring. I call it magical on account of the mysterious manner in which it comes into the world, with the huge shoulder of the mountain rising over it, as if to protect the secret. From under the mountain it silently rises, without visible movement, filling a small natural basin with the stillest blue water. The contrast between the stillness of this basin and the agitation of the water directly after it has overflowed, constitutes half the charm of Vaucluse. The violence of the stream when once it has been set loose on the rocks is as fascinating and indescribable as that of other cataracts; and the rocks in the bed of the Sorgues have been arranged by a master-hand. The setting of the phenomenon struck me as so simple and so fine the vast sad cliff, covered with the afternoon light, still and solid forever, while the liquid element rages and roars at its base that I had no difficulty in understanding the celebrity of Vaucluse. I understood it, but I will not say that I understood Petrarch. He must have been very self-supporting, and Madonna Laura must indeed have been much to him.

The aridity of the hills that shut in the valley is complete, and the whole impression is best conveyed by that very expressive French epithet *morne*. There are the very fragmentary ruins of a castle (of one of the bishops of Cavaillon) on a high spur of the mountain, above the river; and there is another remnant of a feudal habitation on one of the more accessible ledges. Having half an hour to spare before my omnibus was to leave (I must beg the reader's pardon for this atrociously false note; call the vehicle a *diligence*, and for some undiscoverable reason the offence is minimized), I clambered up to this latter spot, and sat among the rocks in the company of a few stunted olives. The Sorgues, beneath me, reaching the plain, flung itself crookedly across the meadows, like an unrolled blue ribbon. I tried to think of the *amant de Laure*, for literature's sake; but I had no great success, and the most I could, do was to say to myself that I must try again. Several months have elapsed since then, and I am ashamed to confess that the trial has not yet come off. The only very definite conviction I arrived at was that Vaucluse is indeed cockneyfied, but that I should have been a fool, all the same, not to come.

36

I mounted into my diligence at the door of the Hotel de Petrarque et de Laure, and we made our way back to Isle-sur-Sorgues in the fading light. This village, where at six o'clock every one appeared to have gone to bed, was fairly darkened by its high, dense plane-trees, under which the rushing river, on a level with its parapets, looked unnaturally, almost wickedly blue. It was a glimpse which has left a picture in my mind: the little closed houses, the place empty and soundless in the autumn dusk but for the noise of waters, and in the middle, amid the blackness of the shade, the gleam of the swift, strange tide. At the station every one was talking of the inundation being in many places an accomplished fact, and, in particular, of the condition of the Durance at some point that I have forgotten. At Avignon, an hour later, I found the water in some of the streets. The sky cleared in the evening, the moon lighted up the submerged suburbs, and the population again collected in the high places to enjoy the spectacle. It exhibited a certain sameness, however, and by nine o'clock there was considerable animation in the Place Crillon, where there is nothing to be seen but the front of the theatre and of several cafes in addition, indeed, to a statue of this celebrated brave, whose valor redeemed some of the numerous military disasters of the reign of Louis XV. The next morning the lower quarters of the town were in a pitiful state; the situation seemed to me odious. To express my disapproval of it, I lost no time in taking the train for Orange, which, with its other attractions, had the merit of not being seated on the Rhone. It was my destiny to move northward; but even if I had been at liberty to follow a less unnatural course I should not then have undertaken it, inasmuch, as the railway between Avignon and Marseilles was credibly reported to be (in places) under water. This was the case with almost everything but the line itself, on the way to Orange. The day proved splendid, and its brilliancy only lighted up the desolation. Farmhouses and cottages were up to their middle in the yellow liquidity; haystacks looked like dull little islands; windows and doors gaped open, without faces; and interruption and flight were represented in the scene. It was brought home to me that the *populations rurales* have many different ways of suffering, and my heart glowed with a grateful sense of cockneyism. It was under the influence of this emotion that I alighted at Orange, to visit a collection of eminently civil monuments.

The collection consists of but two objects, but these objects are so fine that I will let the word pass. One of them is a triumphal arch, supposedly of the period of Marcus Aurelius; the other is a fragment, magnificent in its ruin, of a Roman theatre. But for these fine Roman remains and for its name, Orange is a perfectly featureless little town; without the Rhone which, as I have mentioned, is several miles distant to help it to a physiognomy. It seems one of the oddest things that this obscure French borough obscure, I mean, in our modern era, for the Gallo-Roman Arausio must have been, judging it by its arches and theatre, a place of some importance should have given its name to the heirs apparent of the throne of Holland, and been borne by a king of England who had sovereign rights over it. During the Middle Ages it formed part of an independent principality; but in 1531 it fell, by the marriage of one of its princesses, who had inherited it, into the family of Nassau. I read in my indispensable Murray that it was made over to France by the treaty of Utrecht. The arch of triumph, which stands a little way out of the town, is rather a pretty than an imposing vestige of the Romans. If it had greater purity of style, one might say of it that it belonged to the same family of monuments as the Maison Carree at Nimes. It has three passages, the middle much higher than the others, and a very elevated attic. The vaults of the passages are richly sculptured, and the whole monument is covered with friezes and military trophies. This sculpture is rather mixed; much of it is broken and defaced, and the rest seemed to me ugly, though its workmanship is praised. The arch

144

is at once well preserved and much injured. Its general mass is there, and as Roman monuments go it is remarkably perfect; but it has suffered, in patches, from the extremity of restoration. It is not, on the whole, of absorbing interest. It has a charm, nevertheless, which comes partly from its soft, bright yellow color, partly from a certain elegance of shape, of expression; and on that well-washed Sunday morning, with its brilliant tone, surrounded by its circle of thin poplars, with the green country lying beyond it and a low blue horizon showing through its empty portals, it made, very sufficiently, a picture that hangs itself to one of the lateral hooks of the memory. I can take down the modest composition, and place it before me as I write. I see the shallow, shining puddles in the hard, fair French road; the pale blue sky, diluted by days of rain; the disgarnished autumnal fields; the mild sparkle of the low horizon; the solitary figure in sabots, with a bundle under its arm, advancing along the *chaussee*; and in the middle I see the little ochre-colored monument, which, in spite of its antiquity, looks bright and gay, as everything must look in France of a fresh Sunday morning.

It is true that this was not exactly the appearance of the Roman theatre, which lies on the other side of the town; a fact that did not prevent me from making my way to it in less than five minutes, through a succession of little streets concerning which I have no observations to record. None of the Roman remains in the south of France are more impressive than this stupendous fragment. An enormous mound rises above the place, which was formerly occupied I quote from Murray first by a citadel of the Romans, then by a castle of the princes of Nassau, razed by Louis XIV. Facing this hill a mighty wall erects itself, thirty-six meters high, and composed of massive blocks of dark brown stone, simply laid one on the other; the whole naked, rugged surface of which suggests a natural cliff (say of the Vaucluse order) rather than an effort of human, or even of Roman labor. It is the biggest thing at Orange, it is bigger than all Orange put together, and its permanent massiveness makes light of the shrunken city. The face it presents to the town the top of it garnished with two rows of brackets, perforated with holes to receive the staves of the *velarium* bears the traces of more than one tier of ornamental arches; though how these flat arches were applied, or incrusted, upon the wall, I do not profess to explain. You pass through a diminutive postern which seems in proportion about as high as the entrance of a rabbit-hutch into the lodge of the custodian, who introduces you to the interior of the theatre. Here the mass of the hill affronts you, which the ingenious Romans treated

145

simply as the material of their auditorium. They inserted their stone seats, in a semicircle, in the slope of the lull, and planted their colossal wall opposite to it. This wall, from the inside, is, if possible, even more imposing. It formed the back of the stage, the permanent scene, and its enormous face was coated with marble. It contains three doors, the middle one being the highest, and having above it, far aloft, a deep niche, apparently intended for an imperial statue. A few of the benches remain on the hillside which, however, is mainly a confusion of fragments. There is part of a corridor built into the hill, high up, and on the crest are the remnants of the demolished castle. The whole place is a kind of wilderness of ruin; there are scarcely any details; the great feature is the overtopping wall. This wall being the back of the scene, the space left between it and the chord of the semicircle (of the auditorium) which formed the proscenium is rather less than one would have supposed. In other words, the stage was very shallow, and appears to have been arranged for a number of performers standing in a line, like a company of soldiers. There stands the silent skeleton, however, as impressive by what it leaves you to guess and wonder about as by what it tells you. It has not the sweetness, the softness of melancholy, of the theatre at Arles; but it is more extraordinary, and one can imagine only tremendous tragedies being enacted there, -

"Presenting Thebes' or Pelops' line."

At either end of the stage, coming forward, is an immense wing, immense in height, I mean, as it reaches to the top of the scenic wall; the other dimensions are not remarkable. The division to the right, as you face the stage, is pointed out as the greenroom; its portentous attitude and the open arches at the top give it the air of a well. The compartment on the left is exactly similar, save that it opens into the traces of other chambers, said to be those of a hippodrome adjacent to the theatre. Various fragments are visible which refer themselves plausibly to such an establishment; the greater axis of the hippodrome would appear to have been on a line with the triumphal arch. This is all I saw, and all there was to see, of Orange, which had a very rustic, bucolic aspect, and where I was not even called upon to demand breakfast at the hotel. The entrance of this resort might have been that of a stable of the Roman days.

37

I have been trying to remember whether I fasted all the way to Macon, which I reached at an advanced hour of the evening, and think I must have done so except for the purchase of a box of nougat at Montelimart (the place is famous for the manufacture of this confection, which, at the station, is hawked at the windows of the train) and for a bouillon, very much later, at Lyons. The journey beside the Rhone past Valence, past Tournon, past Vienne would have been charming, on that luminous Sunday, but for two disagreeable accidents. The express from Marseilles, which I took at Orange, was full to overflowing; and the only refuge I could find was an inside angle in a carriage laden with Germans, who had command of the windows, which they occupied as strongly as they have been known to occupy other strategical positions. I scarcely know, however, why I linger on this particular discomfort, for it was but a single item in a considerable list of grievances, grievances dispersed through six weeks of constant railway travel in France. I have not touched upon them at an earlier stage of this chronicle, but my reserve is not owing to any sweetness of association. This form of locomotion, in the country of the amenities, is attended with a dozen discomforts; almost all the conditions of the business are detestable. They force the sentimental tourist again and again to ask himself whether, in consideration of such mortal annoyances, the game is worth the candle. Fortunately, a railway journey is a good deal like a sea voyage; its miseries fade from the mind as soon as you arrive. That is why I completed, to my great satisfaction, my little tour in France. Let this small effusion of ill-nature be my first and last tribute to the whole despotic *gare*: the deadly *salle d'attente*, the insufferable delays over one's luggage, the porterless platform, the overcrowded and illiberal train. How many a time did I permit myself the secret reflection that it is in perfidious Albion that they order this matter best! How many a time did the eager British mercenary, clad in velveteen and clinging to the door of the carriage as it glides into the station, revisit my invidious dreams! The paternal porter and the responsive hansom are among the best gifts of the English genius to the world. I hasten to add, faithful to my habit (so insufferable to some of my friends) of ever and again readjusting the balance after I have given it an honest tip, that the bouillon at Lyons, which I spoke of above, was, though by no means an ideal bouillon, much better than any I could have obtained at an English railway station. After I had imbibed it, I sat in the train (which waited a long time at Lyons) and, by the light of one of the big lamps on the platform, read all sorts of disagreeable things in certain radical newspapers which I had bought at the book-stall. I gathered from these sheets that Lyons was in extreme commotion. The Rhone and the Saone, which form a girdle for the splendid town, were almost in the streets, as I could easily believe from what I had seen of the country after leaving Orange. The Rhone, all the way to Lyons, had been in all sorts of places where it had no business to be, and matters were naturally not improved by its confluence with the charming and copious stream which, at Macon, is said once to have given such a happy opportunity to the egotism of the capital. A visitor from Paris (the anecdote is very old), being asked on the quay of that city whether he didn't admire the Saone, replied good-naturedly that it was very pretty, but that in Paris they spelled it with the *ei*. This moment of general alarm at Lyons had been chosen by certain ingenious persons (I credit them, perhaps, with too sure a prevision of the rise of the rivers) for practicing further upon the apprehensions of the public. A bombshell filled with dynamite had been thrown into a cafe, and various votaries of the comparatively innocuous *petit verre* had been wounded (I am not sure whether any one had been killed) by the irruption. Of course there had been arrests and incarcerations, and the "Intransigeant" and the "Rappel" were filled with the echoes of the explosion. The tone of these organs is rarely edifying, and it had never been less so than on this occasion. I wondered, as I looked through them, whether I was losing all my radicalism; and then I wondered whether, after all, I had any to lose. Even in so long await as that tiresome delay at Lyons I failed to settle the question, any more than I made up my mind as to the probable future of the militant democracy, or the ultimate form of a civilization which should have blown up everything else. A few days later, the waters went down it Lyons; but the democracy has not gone down.

I remember vividly the remainder of that evening which I spent at Macon, remember it with a chattering of the teeth. I know not what had got into the place; the temperature, for the last day of October, was eccentric and incredible. These epithets may also be applied to the hotel itself, an extraordinary structure, all facade, which exposes an uncovered rear to the gaze of nature. There is a demonstrative, voluble landlady, who is of course part of the facade; but everything behind her is a trap for the winds, with chambers, corridors, staircases, all exhibited to the sky, as if the outer wall of the house had been lifted off. It would have been delightful for Florida, but it didn't do for Burgundy, even on the eve of November 1st, so that I suffered absurdly from the rigor of a season that had not yet begun. There was something in the air; I felt it the next day, even on the sunny quay of the Saone, where in spite of a fine southerly exposure I extracted little warmth from the reflection that Alphonse de Lamartine had often trodden the flags. Macon struck me, somehow, as suffering from a chronic numbness, and there was nothing exceptionally cheerful in the remarkable extension of the river. It was no longer a river, it had become a lake; and from my window, in the painted face of the inn, I saw that the opposite bank had been moved back, as it were, indefinitely. Unfortunately, the various objects with which it was furnished had not been moved as well, the consequence of which was an extraordinary confusion in the relations of thing. There were always poplars to be seen, but the poplar had become an aquatic plant. Such phenomena, however, at Macon attract but little attention, as the Saone, at certain seasons of the year, is nothing if not expansive. The people are as used to it as they appeared to be to the bronze statue of Lamartine, which is the principal monument of the *place*, and which, representing the poet in a frogged overcoat and topboots, improvising in a high wind, struck me as even less casual in its attitude than monumental sculpture usually succeeds in being. It is true that in its present position I thought better of this work of art, which is from the hand of M. Falquiere, than when I had seen it through the factitious medium of the Salon of 1876. I walked up the hill where the older part of Macon lies, in search of the natal house of the *amant d'Elvire*, the Petrarch whose Vaucluse was the bosom of the public. The Guide-Joanne quotes from "Les Confidences" a description of the birthplace of the poet, whose treatment of the locality is indeed poetical. It tallies strangely little with the reality, either as regards position or other features; and it may be said to be, not an aid, but a direct obstacle, to a discovery of the house. A very humble edifice, in a small back street, is designated by a municipal tablet, set into its face, as the scene of Lamartine's advent into the world. He himself speaks of a vast and lofty structure, at the angle of a *place*, adorned with iron clamps, with a *porte haute et large* and many other peculiarities. The house with the tablet has two meager stories above the basement, and (at present, at least) an air of extreme shabbiness; the *place*, moreover, never can have been vast. Lamartine was accused of writing history incorrectly, and apparently he started wrong at first: it had never become clear to him where he was born. Or is the tablet wrong? If the house is small, the tablet is very big.

149

38

The foregoing reflections occur, in a cruder form, as it were, in my note-book, where I find this remark appended to them: "Don't take leave of Lamartine on that contemptuous note; it will be easy to think of something more sympathetic!" Those friends of mine, mentioned a little while since, who accuse me of always tipping back the balance, could not desire a paragraph more characteristic; but I wish to give no further evidence of such infirmities, and will therefore hurry away from the subject, hurry away in the train which, very early on a crisp, bright morning, conveyed. me, by way of an excursion, to the ancient city of Bourg-en-Bresse. Shining in early light, the Saone was spread, like a smooth, white tablecloth, over a considerable part of the flat country that I traversed. There is no provision made in this image for the long, transparent screens of thin-twigged trees which rose at intervals out of the watery plain; but as, under the circumstances, there seemed to be no provision for them in fact, I will let my metaphor go for what it is worth. My journey was (as I remember it) of about an hour and a half; but I passed no object of interest, as the phrase is, whatever. The phrase hardly applies even to Bourg itself, which is simply a town *quelconque*, as M. Zola would say. Small, peaceful, rustic, it stands in the midst of the great dairy-feeding plains of Bresse, of which fat county, sometime property of the house of Savoy, it was the modest capital. The blue masses of the Jura give it a creditable horizon, but the only nearer feature it can point to is its famous sepulchral church. This edifice lies at a fortunate distance from the town, which, though inoffensive, is of too common a stamp to consort with such a treasure. All I ever knew of the church of Brou I had gathered, years ago, from Matthew Arnold's beautiful poem, which bears its name. I remember thinking, in those years, that it was impossible verses could be more touching than these; and as I stood before the object of my pilgrimage, in the gay French light (though the place was so dull), I recalled the spot where I had first read them, and where I read them again and yet again, wondering whether it would ever be my fortune to visit the church of Brou. The spot in question was an armchair in a window which looked out on some cows in a field; and whenever I glanced at the cows it came over me I scarcely know why that I should probably never behold the structure reared by the Duchess Margaret. Some of our visions never come to pass; but we must be just, others do. "So sleep, forever sleep, O princely pair!" I remembered that line of Matthew Arnold's, and the stanza about the Duchess Margaret coming to watch the builders on her palfry white. Then there came to me something in regard to the moon shining on winter nights through the cold clere-story. The tone of the place at that hour was not at all lunar; it was cold and bright, but with the chill of an autumn morning; yet this, even with the fact of the unexpected remoteness of the church from the Jura added to it, did not prevent me from feeling that I looked at a monument in the production of which or at least in the effect of which on the tourist mind of to-day Matthew Arnold had been much concerned. By a pardonable license he has placed it a few miles nearer to the forests of the Jura than it stands at present. It is very true that, though the mountains in the sixteenth century can hardly have been in a different position, the plain which separates the church from them may have been bedecked with woods. The visitor to-day cannot help wondering why the beautiful building, with its splendid works of art, is dropped down in that particular spot, which looks so accidental and arbitrary. But there are reasons for most things, and there were reasons why the church of Brou should be at Brou, which is a vague little suburb of a vague little town.

The responsibility rests, at any rate, upon the Duchess Margaret, Margaret of Austria, daughter of the Emperor Maximilian and his wife Mary of Burgundy, daughter of Charles the Bold. This lady has a high name in history, having been regent of the Netherlands in behalf of her nephew, the Emperor Charles V., of whose early education she had had the care. She married in 1501 Philibert the Handsome, Duke of Savoy, to whom the province of Bresse belonged, and who died two years later. She had been betrothed, is a child, to Charles VIII. of France, and was kept for some time at the French court, that of her prospective father-in-law, Louis XI.; but she was eventually repudiated, in order that her *fiance* might marry Anne of Brittany, an alliance so magnificently political that we almost condone the offence to a sensitive princess. Margaret did not want for husbands, however, inasmuch as before her marriage to Philibert she had been united to John of Castile, son of Ferdinand V., King of Aragon, an episode terminated, by the death of the Spanish prince, within a year. She was twenty-two years regent of the Netherlands, and died at fifty-one, in 1530. She might have been, had she chosen, the wife, of Henry VII. of England. She was one of the signers of the League of Cambray, against the Venetian republic, and was a most

151

politic, accomplished, and judicious princess. She undertook to build the church of Brou as a mausoleum, for her second husband and herself, in fulfillment of a vow made by Margaret of Bourbon, mother of Philibert, who died before she could redeem her pledge, and who bequeathed the duty to her son. He died shortly afterwards, and his widow assumed the pious task. According to Murray, she entrusted the erection of the church to "Maistre Loys von Berghem," and the sculpture to "Maistre Conrad." The author of a superstitious but carefully prepared little Notice, which I bought at Bourg, calls the architect and sculptor (at once) Jehan de Paris, author (sic) of the tomb of Francis II. of Brittany, to which we gave some attention at Nantes, and which the writer of my pamphlet ascribes only subordinately to Michel Colomb. The church, which is not of great size, is in the last and most flamboyant phase of Gothic, and in admirable preservation; the west front, before which a quaint old sun-dial is laid out on the ground, a circle of numbers marked in stone, like those on a clock face, let into the earth, is covered with delicate ornament. The great feature, however (the nave is perfectly bare and wonderfully new-looking, though the warden, a stolid yet sharp old peasant, in a blouse, who looked more as if his line were chaffering over turnips than showing off works of art, told me that it has never been touched, and that its freshness is simply the quality of the stone), the great feature is the admirable choir, in the midst of which the three monuments have bloomed under the chisel, like exotic plants in a conservatory. I saw the place to small advantage, for the stained glass of the windows, which are fine, was under repair, and much of it was masked with planks.

In the centre lies Philibert-le-Bel, a figure of white marble on a great slab of black, in his robes and his armor, with two boy-angels holding a tablet at his head, and two more at his feet. On either side of him is another cherub: one guarding his helmet, the other his stiff gauntlets. The attitudes of these charming children, whose faces are all bent upon him in pity, have the prettiest tenderness and respect. The table on which he lies is supported by elaborate columns, adorned with niches containing little images, and with every other imaginable elegance; and beneath it he is represented in that other form, so common in the tombs of the Renaissance, a man naked and dying, with none of the state and splendor of the image above. One of these figures embodies the duke the other simply the mortal; and there is something very strange and striking in the effect of the latter, seen dimly and with difficulty through the intervals of the rich supports of the upper slab. The monument of Margaret herself is on the left, all in white merble, tormented into a multitude of exquisite patterns, the last extravagance of a Gothic which had gone so far that nothing was left it but to return upon itself. Unlike her husband, who has only the high roof of the church above him, she lies under a canopy supported and covered by a wilderness of embroidery, flowers, devices, initials, arabesques, statuettes. Watched over by cherubs, she is also in her robes and ermine, with a greyhound sleeping at her feet (her husband, at his, has a waking lion); and the artist has not, it is to be presumed, represented her as more beautiful than she was. She looks, indeed, like the regent of a turbulent realm. Beneath her couch is stretched another figure, a less brilliant Margaret, wrapped in her shroud, with her long hair over her shoulders. Round the tomb is the battered iron railing placed there originally, with the mysterious motto of the duchess worked into the top, *fortune infortune fortune*. The other two monuments are protected by barriers of the same pattern. That of Margaret of Bourbon, Philibert's mother, stands on the right of the choir; and I suppose its greatest distinction is that it should have been erected to a mother-in-law. It is but little less florid and sumptuous than the others; it has, however, no second recumbent figure. On the other hand, the statuettes that surround the base of the tomb are of even more exquisite workmanship: they represent weeping women, in long mantles and hoods, which latter hang forward over the small face of the figure, giving the artist a chance to carve the features within this hollow of drapery, an extraordinary play of skill. There is a high, white marble shrine of the Virgin, as extraordinary as all the rest (a series of compartments, representing the various scenes of her life, with the Assumption in the middle); and there is a magnificent series of stalls, which are simply the intricate embroidery of the tombs translated into polished oak. All these things are splendid, ingenious, elaborate, precious; it is goldsmith's work on a monumental scale, and the general effect is none the less beautiful and solemn because it is so rich. But the monuments of the church of Brou are not the noblest that one may see; the great tombs of Verona are finer, and various other early Italian work. These things are not insincere, as Ruskin would say; but they are pretentious, and they are not positively *naïfs*. I should mention that the walls of the choir are embroidered in places with Margaret's tantalizing device, which partly, perhaps, because it is tantalizing is so very decorative, as they say in London. I know not whether she was acquainted with this epithet; but she had anticipated one of the fashions most characteristic of our age.

One asks one's self how all this decoration, this luxury of fair and chiseled marble, survived the French Revolution. An hour of liberty in the choir of Brou would have been a carnival for the imagebreakers. The well-fed Bressois are surely a good-natured people. I call them well-fed both on general and on particular grounds. Their province has the most savory aroma, and I found an opportunity to test its reputation. I walked back into the town from the church (there was really nothing to be seen by the way), and as the hour of the midday breakfast had struck, directed my steps to the inn. The table d'hote was going on, and a gracious, bustling, talkative landlady welcomed me. I had an excellent repast the best repast possible which consisted simply of boiled eggs and bread and butter. It was the quality of these simple ingredients that made the occasion memorable. The eggs were so good that I am ashamed to say how many of them I consumed. "La plus belle fille du monde," as the French proverb says, "ne peut donner que ce qu'elle a;" and it might seem that an egg which has succeeded in being fresh has done all that can reasonably be expected of it. But there was a bloom of punctuality, so to speak, about these eggs of Bourg, as if it had been the intention of the very hens themselves that they should be promptly served. "Nous sommes en Bresse, et le beurre n'est pas mauvais," the landlady said, with a sort of dry coquetry, as she placed this article before me. It was the poetry of butter, and I ate a pound or two of it; after which I came away with a strange mixture of impressions of late Gothic sculpture and thick *tartines*. I came away through the town, where, on a little green promenade, facing the hotel, is a bronze statue of Bichat, the physiologist, who was a Bressois. I mention it, not on account of its merit (though, as statues go, I don't remember that it is bad), but because I learned from it my ignorance, doubtless, did me little honor that Bichat had died at thirty years of age, and this revelation was almost agitating. To have done so much in so short a life was to be truly great. This reflection, which looks deplorably trite as I write it here, had the effect of eloquence as I uttered it, for my own benefit, on the bare little mall at Bourg.

39

On my return to Macon I found myself fairly face to face with the fact that my little tour was near its end. Dijon had been marked by fate as its farthest limit, and Dijon was close at hand. After that I was to drop the tourist, and re-enter Paris as much as possible like a Parisian. Out of Paris the Parisian never loiters, and therefore it would be impossible for me to stop between Dijon and the capital. But I might be a tourist a few hours longer by stopping somewhere between Macon and Dijon. The question was where I should spend these hours. Where better, I asked myself (for reasons not now entirely clear to me) than at Beaune? On my way to this town I passed the stretch of the Cote d'Or, which, covered with a mellow autumn haze, with the sunshine shimmering through, looked indeed like a golden slope. One regards with a kind of awe the region in which the famous *crus* of Burgundy (Yougeot, Chambertin, Nuits, Beaune) are, I was going to say, manufactured. Adieu, paniers; vendanges sont faites! The vintage was over; the shrunken russet fibers alone clung to their ugly stick. The horizon on the left of the road had a charm, however, there is something picturesque in the big, comfortable shoulders of the Cote. That delicate critic, M. Emile Montegut, in a charming record of travel through this region, published some years ago, praises Shakespeare for having talked (in "Lear") of "waterish Burgundy." Vinous Burgundy would surely be more to the point. I stopped at Beaune in pursuit of the picturesque, but I might almost have seen the little I discovered without stopping. It is a drowsy little Burgundian town, very old and ripe, with crooked streets, vistas always oblique, and steep, moss-covered roofs. The principal lion is the Hopital-Saint-Esprit, or the Hotel-Dieu, simply, as they call it there, founded in 1443 by Nicholas Rollin, Chancellor of Burgundy. It is administered by the sisterhood of the Holy Ghost, and is one of the most venerable and stately of hospitals. The face it presents to the street is simple, but striking, a plain, windowless wall, surmounted by a vast slate roof, of almost mountainous steepness. Astride this roof sits a tall, slate-covered spire, from which, as I arrived, the prettiest chimes I ever heard (worse luck to them, as I will presently explain) were ringing. Over the door is a high, quaint canopy, without supports, with its vault painted blue and covered with gilded stars. (This, and indeed the whole building, have lately been restored, and its antiquity is quite of the spick-and-span order.

But it is very delightful.) The treasure of the place is a precious picture, a Last Judgment, attributed equally to John van Eyck and Roger van der Weyden, given to the hospital in the fifteenth century by Nicholas Rollin aforesaid.

I learned, however, to my dismay, from a sympathizing but inexorable concierge, that what remained to me of the time I had to spend at Beaune, between trains, I had rashly wasted half an hour of it in breakfasting at the station, was the one hour of the day (that of the dinner of the nuns; the picture is in their refectory) during which the treasure could not be shown. The purpose of the musical chimes to which I had so artlessly listened was to usher in this fruitless interval. The regulation was absolute, and my disappointment relative, as I have been happy to reflect since I "looked up" the picture. Crowe and Cavalcaselle assign it without hesitation to Roger van der Weyden, and give a weak little drawing of it in their "Flemish Painters." I learn from them also what I was ignorant of that Nicholas Ronin, Chancellor of Burgundy and founder of the establishment at Beaune, was the original of the worthy kneeling before the Virgin, in the magnificent John van Eyck of the Salon Carre. All I could see was the court of the hospital and two or three rooms. The court, with its tall roofs, its pointed gables and spires, its wooden galleries, its ancient well, with an elaborate superstructure of wrought iron, is one of those places into which a sketcher ought to be let loose. It looked Flemish or English rather than French, and a splendid tidiness pervaded it. The porter took me into two rooms on the ground-floor, into which the sketcher should also be allowed to penetrate; for they made irresistible pictures. One of them, of great proportions, painted in elaborate "subjects," like a ball-room of the seventeenth century, was filled with the beds of patients, all draped in curtains of dark red cloth, the traditional uniform of these, eleemosynary couches. Among them the sisters moved about, in their robes of white flannel, with big white linen hoods. The other room was a strange, immense apartment, lately restored with much splendor. It was of great length and height, had a painted and gilded barrel-roof, and one end of it the one I was introduced to appeared to serve as a chapel, as two white-robed sisters were on their knees before an altar. This was divided by red curtains from the larger part; but the porter lifted one of the curtains, and showed me that the rest of it, a long, imposing vista, served as a ward, lined with little red-draped beds. "C'est l'heure de la lecture," remarked my guide; and a group of convalescents all the patients I saw were women were gathered in the centre around a nun, the points of whose white hood nodded a little above them, and whose gentle voice came to us faintly, with a little echo, down the high perspective. I know not what the good sister was reading, a dull book, I am afraid, but there was so much color, and such a fine, rich air of tradition about the whole place, that it seemed to me I would have risked listening to her. I turned away, however, with that sense of defeat which is always irritating to the appreciative tourist, and pottered about Beaune rather vaguely for the rest of my hour: looked at the statue of Gaspard Monge, the mathematician, in the little *place* (there is no *place* in France too little to contain an effigy to a glorious son); at the fine old porch completely despoiled at the Revolution of the principal church; and even at the meager treasures of a courageous but melancholy little museum, which has been arranged part of it being the gift of a local collector in a small hotel de ville. I carried away from Beaune the impression of something mildly autumnal, something rusty yet kindly, like the taste of a sweet russet pear.

40

It was very well that my little tour was to terminate at Dijon; for I found, rather to my chagrin, that there was not a great deal, from the pictorial point of view, to be done with Dijon. It was no great matter, for I held my proposition to have been by this time abundantly demonstrated, the proposition with which I started: that if Paris is France, France is by no means Paris. If Dijon was a good deal of a disappointment, I felt, therefore, that I could afford it. It was time for me to reflect, also, that for my disappointments, as a general thing, I had only myself to thank. They had too often been the consequence of arbitrary preconceptions, produced by influences of which I had lost the trace. At any rate, I will say plumply that the ancient capital of Burgundy is wanting in character; it is not up to the mark. It is old and narrow and crooked, and it has been left pretty well to itself: but it is not high and overhanging; it is not, to the eye, what the Burgundian capital should be. It has some tortuous vistas, some mossy roofs, some bulging fronts, some gray-faced hotels, which look as if in former centuries in the last, for instance, during the time of that delightful President de Brosses, whose Letters from Italy throw an interesting side-light on Dijon they had witnessed a considerable amount of good living. But there is nothing else. I speak as a man who for some reason which he doesn't remember now, did not pay a visit to the celebrated Puits de Moise, an ancient cistern, embellished with a sculptured figure of the Hebrew lawgiver.

The ancient palace of the Dukes of Burgundy, long since converted into an hotel de ville, presents to a wide, clean court, paved with washed-looking stones, and to a small semicircular *place*, opposite, which looks as if it had tried to be symmetrical and had failed, a facade and two wings, characterized by the stiffness, but not by the grand air, of the early part of the eighteenth century. It contains, however, a large and rich museum, a museum really worthy of a capital. The gem of this exhibition is the great banqueting-hall of the old palace, one of the few features of the place that has not been essentially altered. Of great height, roofed with the old beams and cornices, it contains, filling one end, a colossal Gothic chimneypiece, with a fireplace large enough to roast, not an ox, but a herd of oxen. In the middle of this striking hall, the walls of which. are covered with objects more or less precious, have been placed the tombs of Philippe-le-Hardi and Jean-sans-Peur. These monuments, very splendid in their general effect, have a limited interest. The limitation comes from the fact that we see them to-day in a transplanted and mutilated condition. Placed originally in a church which has disappeared from the face of the earth, demolished and dispersed at the Revolution, they have been reconstructed and restored out of fragments recovered and pieced together. The piecing his been beautifully done; it is covered with gilt and with brilliant paint; the whole result is most artistic. But the spell of the old mortuary figures is broken, and it will never work again. Meanwhile the monuments are immensely decorative.

I think the thing that pleased me best at Dijon was the little old Parc, a charming public garden, about a mile from the town, to which I walked by a long, straight autumnal avenue. It is a *jardin francais* of the last century, a dear old place, with little blue-green perspectives and alleys and *rondpoints*, in which everything balances. I went there late in the afternoon, without meeting a creature, though I had hoped I should meet the President de Brosses. At the end of it was a little river that looked like a canal, and on the further bank was an old-fashioned villa, close to the water, with a little French garden of its own. On the hither side was a bench, on which I seated myself, lingering a good while; for this was just the sort of place I like. It was the furthermost point of my little tour. I thought that over, as I sat there, on the eve of taking the express to Paris; and as the light faded in the Parc the vision of some of the things I had seen became more distinct.

www.ingramcontent.com/pod-product-compliance
Lightning Source LLC
Chambersburg PA
CBHW060032210326
41520CB00009B/1103